Learn the Addition & Subtraction Facts

by

Stacy Otillio & Frank Otillio

CLAY MAZE.com

EMPOWERING CHILDREN
FOR A SUCCESSFUL FUTURE

TABLE OF CONTENTS

TABLE OF CONTENTS

SECTION 1

Addition Facts Sums up to 10

11 worksheets

20 problems per sheet

This section focuses only on the lower addition facts with sums up to 10.

Name ____________________ Date ____________________

1 Addition Facts Sums up to 10

Add.

1) 1 +0	2) 5 +5	3) 2 +3	4) 3 +0	5) 3 +4
6) 7 +1	7) 2 +5	8) 4 +2	9) 4 +5	10) 0 +4
11) 3 +1	12) 6 +4	13) 4 +3	14) 2 +8	15) 3 +3
16) 2 +6	17) 3 +7	18) 7 +3	19) 6 +3	20) 3 +5

Name ______________________ Date ______________________

1 Addition Facts Sums up to 10

Add.

1) $\begin{array}{r} 8 \\ +2 \\ \hline \end{array}$	2) $\begin{array}{r} 5 \\ +1 \\ \hline \end{array}$	3) $\begin{array}{r} 0 \\ +6 \\ \hline \end{array}$	4) $\begin{array}{r} 2 \\ +5 \\ \hline \end{array}$	5) $\begin{array}{r} 4 \\ +1 \\ \hline \end{array}$
6) $\begin{array}{r} 4 \\ +0 \\ \hline \end{array}$	7) $\begin{array}{r} 1 \\ +9 \\ \hline \end{array}$	8) $\begin{array}{r} 0 \\ +2 \\ \hline \end{array}$	9) $\begin{array}{r} 6 \\ +3 \\ \hline \end{array}$	10) $\begin{array}{r} 3 \\ +7 \\ \hline \end{array}$
11) $\begin{array}{r} 4 \\ +3 \\ \hline \end{array}$	12) $\begin{array}{r} 5 \\ +3 \\ \hline \end{array}$	13) $\begin{array}{r} 7 \\ +2 \\ \hline \end{array}$	14) $\begin{array}{r} 7 \\ +3 \\ \hline \end{array}$	15) $\begin{array}{r} 3 \\ +3 \\ \hline \end{array}$
16) $\begin{array}{r} 3 \\ +2 \\ \hline \end{array}$	17) $\begin{array}{r} 5 \\ +4 \\ \hline \end{array}$	18) $\begin{array}{r} 5 \\ +2 \\ \hline \end{array}$	19) $\begin{array}{r} 2 \\ +4 \\ \hline \end{array}$	20) $\begin{array}{r} 2 \\ +7 \\ \hline \end{array}$

Name ______________________ Date ______________________

1 Addition Facts Sums up to 10

Add.

1) $\begin{array}{r} 1 \\ +9 \\ \hline \end{array}$	2) $\begin{array}{r} 4 \\ +1 \\ \hline \end{array}$	3) $\begin{array}{r} 2 \\ +8 \\ \hline \end{array}$	4) $\begin{array}{r} 2 \\ +7 \\ \hline \end{array}$	5) $\begin{array}{r} 3 \\ +7 \\ \hline \end{array}$
6) $\begin{array}{r} 3 \\ +0 \\ \hline \end{array}$	7) $\begin{array}{r} 0 \\ +5 \\ \hline \end{array}$	8) $\begin{array}{r} 0 \\ +8 \\ \hline \end{array}$	9) $\begin{array}{r} 5 \\ +1 \\ \hline \end{array}$	10) $\begin{array}{r} 8 \\ +2 \\ \hline \end{array}$
11) $\begin{array}{r} 1 \\ +4 \\ \hline \end{array}$	12) $\begin{array}{r} 5 \\ +5 \\ \hline \end{array}$	13) $\begin{array}{r} 3 \\ +6 \\ \hline \end{array}$	14) $\begin{array}{r} 5 \\ +3 \\ \hline \end{array}$	15) $\begin{array}{r} 7 \\ +3 \\ \hline \end{array}$
16) $\begin{array}{r} 3 \\ +2 \\ \hline \end{array}$	17) $\begin{array}{r} 5 \\ +2 \\ \hline \end{array}$	18) $\begin{array}{r} 6 \\ +4 \\ \hline \end{array}$	19) $\begin{array}{r} 3 \\ +3 \\ \hline \end{array}$	20) $\begin{array}{r} 6 \\ +3 \\ \hline \end{array}$

Name ______________________ Date ______________________

1 Addition Facts Sums up to 10

Add.

1) 4 +5	2) 8 +1	3) 2 +8	4) 4 +4	5) 3 +5
6) 2 +7	7) 1 +7	8) 3 +6	9) 6 +3	10) 3 +0
11) 5 +0	12) 0 +6	13) 3 +3	14) 1 +2	15) 7 +2
16) 2 +2	17) 5 +1	18) 5 +4	19) 6 +2	20) 2 +6

Name ____________________ Date ____________________

1 Addition Facts Sums up to 10

Add.

1) $\begin{array}{r} 1 \\ +6 \\ \hline \end{array}$	2) $\begin{array}{r} 3 \\ +2 \\ \hline \end{array}$	3) $\begin{array}{r} 2 \\ +3 \\ \hline \end{array}$	4) $\begin{array}{r} 7 \\ +0 \\ \hline \end{array}$	5) $\begin{array}{r} 2 \\ +7 \\ \hline \end{array}$
6) $\begin{array}{r} 8 \\ +2 \\ \hline \end{array}$	7) $\begin{array}{r} 3 \\ +7 \\ \hline \end{array}$	8) $\begin{array}{r} 2 \\ +4 \\ \hline \end{array}$	9) $\begin{array}{r} 0 \\ +2 \\ \hline \end{array}$	10) $\begin{array}{r} 1 \\ +9 \\ \hline \end{array}$
11) $\begin{array}{r} 4 \\ +5 \\ \hline \end{array}$	12) $\begin{array}{r} 5 \\ +2 \\ \hline \end{array}$	13) $\begin{array}{r} 5 \\ +4 \\ \hline \end{array}$	14) $\begin{array}{r} 0 \\ +5 \\ \hline \end{array}$	15) $\begin{array}{r} 6 \\ +2 \\ \hline \end{array}$
16) $\begin{array}{r} 1 \\ +7 \\ \hline \end{array}$	17) $\begin{array}{r} 4 \\ +2 \\ \hline \end{array}$	18) $\begin{array}{r} 2 \\ +2 \\ \hline \end{array}$	19) $\begin{array}{r} 3 \\ +3 \\ \hline \end{array}$	20) $\begin{array}{r} 4 \\ +3 \\ \hline \end{array}$

Name ______________________ Date ______________________

1 Addition Facts Sums up to 10

Add.

1) 4 +2	2) 3 +4	3) 3 +0	4) 9 +0	5) 2 +7
6) 4 +6	7) 7 +2	8) 4 +5	9) 6 +2	10) 4 +4
11) 3 +2	12) 3 +3	13) 3 +1	14) 6 +0	15) 5 +4
16) 2 +2	17) 1 +2	18) 4 +3	19) 3 +5	20) 1 +8

Name ______________________ Date ______________________

1 Addition Facts Sums up to 10

Add.

1) 0 +8	2) 1 +9	3) 4 +3	4) 3 +1	5) 2 +2
6) 4 +1	7) 5 +4	8) 0 +5	9) 6 +2	10) 3 +4
11) 3 +2	12) 4 +5	13) 6 +4	14) 4 +4	15) 7 +3
16) 3 +6	17) 2 +3	18) 3 +3	19) 5 +2	20) 3 +5

Name ______________________ Date ______________________

1 Addition Facts Sums up to 10

Add.

1) 2 +4	2) 2 +5	3) 8 +2	4) 1 +6	5) 7 +1
6) 8 +0	7) 1 +7	8) 2 +8	9) 2 +3	10) 3 +0
11) 3 +7	12) 0 +2	13) 4 +6	14) 4 +5	15) 5 +4
16) 2 +6	17) 5 +3	18) 3 +6	19) 5 +2	20) 7 +2

Name ______________________ Date ______________________

1 Addition Facts Sums up to 10

Add.

1) 0 + 8 = ___	2) 7 + 2 = ___	3) 2 + 2 = ___	4) 4 + 4 = ___	5) 6 + 0 = ___
6) 7 + 3 = ___	7) 3 + 2 = ___	8) 1 + 5 = ___	9) 4 + 3 = ___	10) 2 + 0 = ___
11) 2 + 5 = ___	12) 3 + 7 = ___	13) 1 + 4 = ___	14) 2 + 7 = ___	15) 2 + 8 = ___
16) 6 + 2 = ___	17) 3 + 4 = ___	18) 3 + 1 = ___	19) 3 + 3 = ___	20) 6 + 4 = ___

Name ____________________ Date ____________________

1 Addition Facts Sums up to 10

Add.

1) $\begin{array}{r} 3 \\ +5 \\ \hline \end{array}$ 2) $\begin{array}{r} 1 \\ +4 \\ \hline \end{array}$ 3) $\begin{array}{r} 7 \\ +1 \\ \hline \end{array}$ 4) $\begin{array}{r} 4 \\ +5 \\ \hline \end{array}$ 5) $\begin{array}{r} 0 \\ +9 \\ \hline \end{array}$

6) $\begin{array}{r} 5 \\ +2 \\ \hline \end{array}$ 7) $\begin{array}{r} 2 \\ +3 \\ \hline \end{array}$ 8) $\begin{array}{r} 1 \\ +2 \\ \hline \end{array}$ 9) $\begin{array}{r} 4 \\ +3 \\ \hline \end{array}$ 10) $\begin{array}{r} 0 \\ +1 \\ \hline \end{array}$

11) $\begin{array}{r} 2 \\ +5 \\ \hline \end{array}$ 12) $\begin{array}{r} 7 \\ +3 \\ \hline \end{array}$ 13) $\begin{array}{r} 2 \\ +8 \\ \hline \end{array}$ 14) $\begin{array}{r} 2 \\ +2 \\ \hline \end{array}$ 15) $\begin{array}{r} 7 \\ +2 \\ \hline \end{array}$

16) $\begin{array}{r} 0 \\ +3 \\ \hline \end{array}$ 17) $\begin{array}{r} 5 \\ +5 \\ \hline \end{array}$ 18) $\begin{array}{r} 6 \\ +3 \\ \hline \end{array}$ 19) $\begin{array}{r} 3 \\ +6 \\ \hline \end{array}$ 20) $\begin{array}{r} 5 \\ +3 \\ \hline \end{array}$

Name ______________________ Date ______________________

1 Addition Facts Sums up to 10

Add.

1) 1 +3	2) 6 +3	3) 6 +2	4) 3 +3	5) 0 +8
6) 7 +2	7) 0 +5	8) 5 +5	9) 1 +5	10) 3 +5
11) 4 +6	12) 5 +0	13) 8 +2	14) 4 +3	15) 5 +2
16) 5 +4	17) 6 +4	18) 4 +4	19) 2 +3	20) 2 +7

SECTION 2

Missing Addends Sums up to 10

11 worksheets

15 problems per sheet

This fill-in-the-blank section helps with the transition from addition to subtraction and focuses only on the lower addition facts with sums up to 10.

Name ______________________ Date ______________________

2 Missing Addends Sums up to 10

Fill in the blanks with the missing addends.

1) 0 + ____ = 7   2) 0 + ____ = 5   3) 5 + ____ = 7

4) 8 + ____ = 9   5) 0 + ____ = 3   6) 2 + ____ = 9

7) 1 + ____ = 3   8) 4 + ____ = 5   9) 2 + ____ = 7

10) 6 + ____ = 10   11) 6 + ____ = 8   12) 6 + ____ = 9

13) 5 + ____ = 8   14) 2 + ____ = 6   15) 3 + ____ = 9

Name ______________________ Date ______________________

2 Missing Addends Sums up to 10

Fill in the blanks with the missing addends.

1) 1+ ____ =8

2) 1+ ____ =3

3) 3+ ____ =5

4) 9+ ____ =10

5) 0+ ____ =1

6) 2+ ____ =4

7) 3+ ____ =3

8) 2+ ____ =8

9) 6+ ____ =9

10) 5+ ____ =9

11) 4+ ____ =9

12) 3+ ____ =6

13) 7+ ____ =10

14) 6+ ____ =8

15) 3+ ____ =9

Name ______________________ Date ______________________

2 Missing Addends Sums up to 10

Fill in the blanks with the missing addends.

1) $1 + \_\_\_\_ = 2$

2) $2 + \_\_\_\_ = 5$

3) $3 + \_\_\_\_ = 4$

4) $2 + \_\_\_\_ = 6$

5) $6 + \_\_\_\_ = 7$

6) $5 + \_\_\_\_ = 9$

7) $4 + \_\_\_\_ = 6$

8) $6 + \_\_\_\_ = 10$

9) $3 + \_\_\_\_ = 7$

10) $5 + \_\_\_\_ = 8$

11) $0 + \_\_\_\_ = 2$

12) $4 + \_\_\_\_ = 8$

13) $4 + \_\_\_\_ = 7$

14) $7 + \_\_\_\_ = 7$

15) $2 + \_\_\_\_ = 9$

Name ______________________ Date ______________________

2 Missing Addends Sums up to 10

Fill in the blanks with the missing addends.

1) 4 + ____ = 6
2) 2 + ____ = 2
3) 6 + ____ = 9

4) 8 + ____ = 9
5) 5 + ____ = 8
6) 0 + ____ = 5

7) 0 + ____ = 9
8) 5 + ____ = 9
9) 2 + ____ = 9

10) 1 + ____ = 2
11) 6 + ____ = 8
12) 5 + ____ = 7

13) 1 + ____ = 8
14) 5 + ____ = 10
15) 7 + ____ = 10

Name ______________________ Date ______________________

2 Missing Addends Sums up to 10

Fill in the blanks with the missing addends.

1) $2 + \_\_\_\_ = 7$ 2) $3 + \_\_\_\_ = 4$ 3) $7 + \_\_\_\_ = 7$

4) $0 + \_\_\_\_ = 1$ 5) $7 + \_\_\_\_ = 10$ 6) $1 + \_\_\_\_ = 5$

7) $0 + \_\_\_\_ = 3$ 8) $1 + \_\_\_\_ = 3$ 9) $3 + \_\_\_\_ = 10$

10) $6 + \_\_\_\_ = 9$ 11) $2 + \_\_\_\_ = 9$ 12) $2 + \_\_\_\_ = 5$

13) $4 + \_\_\_\_ = 9$ 14) $2 + \_\_\_\_ = 6$ 15) $7 + \_\_\_\_ = 9$

Name ______________________ Date ______________________

2 Missing Addends Sums up to 10

Fill in the blanks with the missing addends.

1) $5 + \_\_\_\_ = 7$	2) $0 + \_\_\_\_ = 6$	3) $8 + \_\_\_\_ = 8$
4) $9 + \_\_\_\_ = 10$	5) $2 + \_\_\_\_ = 4$	6) $1 + \_\_\_\_ = 7$
7) $4 + \_\_\_\_ = 8$	8) $3 + \_\_\_\_ = 6$	9) $3 + \_\_\_\_ = 7$
10) $6 + \_\_\_\_ = 8$	11) $0 + \_\_\_\_ = 1$	12) $1 + \_\_\_\_ = 6$
13) $5 + \_\_\_\_ = 10$	14) $4 + \_\_\_\_ = 7$	15) $5 + \_\_\_\_ = 9$

Name ______________________ Date ______________________

2 Missing Addends Sums up to 10

Fill in the blanks with the missing addends.

1) 3+____=3

2) 0+____=7

3) 3+____=9

4) 1+____=4

5) 4+____=9

6) 8+____=10

7) 6+____=8

8) 7+____=8

9) 6+____=9

10) 1+____=3

11) 4+____=6

12) 3+____=10

13) 2+____=6

14) 3+____=5

15) 3+____=7

Name ______________________ Date ______________________

2 Missing Addends Sums up to 10

Fill in the blanks with the missing addends.

1) $2 + \_\_\_\_ = 7$ 2) $0 + \_\_\_\_ = 9$ 3) $9 + \_\_\_\_ = 10$

4) $1 + \_\_\_\_ = 7$ 5) $5 + \_\_\_\_ = 7$ 6) $5 + \_\_\_\_ = 9$

7) $1 + \_\_\_\_ = 9$ 8) $3 + \_\_\_\_ = 5$ 9) $2 + \_\_\_\_ = 9$

10) $4 + \_\_\_\_ = 6$ 11) $4 + \_\_\_\_ = 9$ 12) $3 + \_\_\_\_ = 6$

13) $4 + \_\_\_\_ = 7$ 14) $5 + \_\_\_\_ = 8$ 15) $7 + \_\_\_\_ = 10$

Name ______________________ Date ______________________

2 Missing Addends Sums up to 10

Fill in the blanks with the missing addends.

1) $2 + \_\_\_\_ = 9$ 2) $7 + \_\_\_\_ = 10$ 3) $3 + \_\_\_\_ = 9$

4) $3 + \_\_\_\_ = 10$ 5) $4 + \_\_\_\_ = 7$ 6) $2 + \_\_\_\_ = 4$

7) $1 + \_\_\_\_ = 6$ 8) $6 + \_\_\_\_ = 9$ 9) $2 + \_\_\_\_ = 10$

10) $3 + \_\_\_\_ = 7$ 11) $8 + \_\_\_\_ = 8$ 12) $0 + \_\_\_\_ = 8$

13) $4 + \_\_\_\_ = 6$ 14) $4 + \_\_\_\_ = 8$ 15) $3 + \_\_\_\_ = 8$

Name ______________________________ Date ______________________________

2 Missing Addends Sums up to 10

Fill in the blanks with the missing addends.

1) 6+____=10 2) 7+____=8 3) 4+____=6

4) 8+____=10 5) 3+____=10 6) 2+____=10

7) 3+____=4 8) 7+____=7 9) 5+____=7

10) 6+____=9 11) 2+____=8 12) 1+____=9

13) 2+____=9 14) 0+____=8 15) 0+____=2

Name ______________________ Date ______________________

2 Missing Addends Sums up to 10

Fill in the blanks with the missing addends.

1) 4+____=9 2) 5+____=8 3) 1+____=1

4) 7+____=10 5) 6+____=6 6) 1+____=8

7) 8+____=10 8) 1+____=5 9) 4+____=10

10) 2+____=4 11) 3+____=9 12) 4+____=7

13) 4+____=8 14) 1+____=7 15) 2+____=9

SECTION 3

Subtraction Facts Minuends up to 10

11 worksheets

20 problems per sheet

This section focuses only on the lower subtraction facts with minuends up to 10.

Name ____________________ Date ____________________

3 Subtraction Facts Minuends up to 10

Subtract.

1) 10 − 7 = ___

2) 9 − 7 = ___

3) 8 − 2 = ___

4) 3 − 3 = ___

5) 7 − 3 = ___

6) 2 − 2 = ___

7) 4 − 2 = ___

8) 10 − 2 = ___

9) 5 − 1 = ___

10) 9 − 1 = ___

11) 9 − 3 = ___

12) 6 − 2 = ___

13) 7 − 2 = ___

14) 6 − 3 = ___

15) 4 − 3 = ___

16) 8 − 6 = ___

17) 10 − 8 = ___

18) 7 − 7 = ___

19) 8 − 4 = ___

20) 10 − 3 = ___

Name ______________________ Date ______________________

3 Subtraction Facts Minuends up to 10

Subtract.

1) $\begin{array}{r} 1 \\ -0 \\ \hline \end{array}$	2) $\begin{array}{r} 10 \\ -7 \\ \hline \end{array}$	3) $\begin{array}{r} 8 \\ -2 \\ \hline \end{array}$	4) $\begin{array}{r} 4 \\ -4 \\ \hline \end{array}$	5) $\begin{array}{r} 8 \\ -4 \\ \hline \end{array}$
6) $\begin{array}{r} 2 \\ -1 \\ \hline \end{array}$	7) $\begin{array}{r} 9 \\ -2 \\ \hline \end{array}$	8) $\begin{array}{r} 9 \\ -7 \\ \hline \end{array}$	9) $\begin{array}{r} 5 \\ -5 \\ \hline \end{array}$	10) $\begin{array}{r} 10 \\ -3 \\ \hline \end{array}$
11) $\begin{array}{r} 9 \\ -6 \\ \hline \end{array}$	12) $\begin{array}{r} 10 \\ -2 \\ \hline \end{array}$	13) $\begin{array}{r} 5 \\ -4 \\ \hline \end{array}$	14) $\begin{array}{r} 10 \\ -5 \\ \hline \end{array}$	15) $\begin{array}{r} 6 \\ -3 \\ \hline \end{array}$
16) $\begin{array}{r} 5 \\ -2 \\ \hline \end{array}$	17) $\begin{array}{r} 4 \\ -2 \\ \hline \end{array}$	18) $\begin{array}{r} 6 \\ -2 \\ \hline \end{array}$	19) $\begin{array}{r} 10 \\ -6 \\ \hline \end{array}$	20) $\begin{array}{r} 8 \\ -6 \\ \hline \end{array}$

Name ________________ Date ________________

3 Subtraction Facts Minuends up to 10

Subtract.

1) 10 − 9	2) 8 − 1	3) 7 − 5	4) 10 − 6	5) 8 − 5
6) 5 − 3	7) 5 − 0	8) 10 − 2	9) 6 − 0	10) 10 − 4
11) 8 − 0	12) 6 − 4	13) 9 − 3	14) 5 − 2	15) 8 − 4
16) 9 − 6	17) 10 − 7	18) 8 − 6	19) 7 − 3	20) 10 − 8

Name ______________________ Date ______________________

3 Subtraction Facts Minuends up to 10

Subtract.

1) $5 - 5$	2) $9 - 5$	3) $10 - 6$	4) $7 - 0$	5) $4 - 1$
6) $8 - 2$	7) $8 - 0$	8) $8 - 6$	9) $5 - 3$	10) $5 - 1$
11) $6 - 3$	12) $3 - 2$	13) $10 - 4$	14) $10 - 7$	15) $9 - 6$
16) $10 - 5$	17) $10 - 2$	18) $9 - 4$	19) $4 - 2$	20) $6 - 2$

Name ______________________ Date ______________________

3 Subtraction Facts Minuends up to 10

Subtract.

1) $7 - 3$	2) $3 - 2$	3) $9 - 5$	4) $10 - 2$	5) $5 - 5$
6) $7 - 4$	7) $4 - 3$	8) $7 - 6$	9) $7 - 0$	10) $6 - 5$
11) $10 - 4$	12) $6 - 4$	13) $10 - 7$	14) $8 - 6$	15) $9 - 7$
16) $5 - 2$	17) $10 - 8$	18) $8 - 2$	19) $8 - 5$	20) $9 - 2$

Name ______________________ Date ______________________

3 Subtraction Facts Minuends up to 10

Subtract.

1) 7 − 7	2) 4 − 2	3) 9 − 2	4) 9 − 7	5) 5 − 0
6) 6 − 6	7) 6 − 2	8) 7 − 4	9) 8 − 6	10) 6 − 1
11) 5 − 3	12) 8 − 5	13) 10 − 4	14) 10 − 6	15) 7 − 5
16) 10 − 1	17) 10 − 9	18) 6 − 5	19) 5 − 2	20) 10 − 3

Name ________________ Date ________________

3 Subtraction Facts Minuends up to 10

Subtract.

1) 6 − 3	2) 8 − 2	3) 8 − 6	4) 4 − 2	5) 1 − 0
6) 8 − 5	7) 7 − 7	8) 9 − 9	9) 10 − 8	10) 10 − 7
11) 4 − 1	12) 10 − 2	13) 9 − 6	14) 8 − 4	15) 10 − 3
16) 5 − 3	17) 5 − 2	18) 7 − 4	19) 10 − 6	20) 9 − 4

Name ______________________ Date ______________________

3 Subtraction Facts Minuends up to 10

Subtract.

1) $8 - 5$ = ____

2) $8 - 4$ = ____

3) $7 - 3$ = ____

4) $5 - 4$ = ____

5) $9 - 9$ = ____

6) $10 - 1$ = ____

7) $4 - 1$ = ____

8) $9 - 5$ = ____

9) $8 - 0$ = ____

10) $7 - 0$ = ____

11) $4 - 2$ = ____

12) $7 - 5$ = ____

13) $7 - 2$ = ____

14) $5 - 2$ = ____

15) $7 - 4$ = ____

16) $6 - 2$ = ____

17) $10 - 2$ = ____

18) $9 - 6$ = ____

19) $10 - 7$ = ____

20) $9 - 4$ = ____

Name ______________________ Date ______________________

3 Subtraction Facts Minuends up to 10

Subtract.

1) 10 − 6	2) 3 − 3	3) 10 − 5	4) 4 − 2	5) 6 − 5
6) 8 − 5	7) 2 − 0	8) 7 − 6	9) 10 − 7	10) 7 − 1
11) 9 − 5	12) 5 − 2	13) 9 − 7	14) 8 − 2	15) 6 − 2
16) 9 − 4	17) 9 − 6	18) 5 − 3	19) 9 − 3	20) 9 − 2

Name ______________________ Date ______________________

3 Subtraction Facts Minuends up to 10

Subtract.

1) 6 − 2	2) 5 − 2	3) 10 − 8	4) 5 − 0	5) 8 − 3
6) 6 − 1	7) 6 − 4	8) 8 − 8	9) 10 − 1	10) 4 − 0
11) 9 − 1	12) 10 − 6	13) 9 − 6	14) 8 − 6	15) 9 − 4
16) 9 − 3	17) 7 − 3	18) 9 − 2	19) 4 − 2	20) 9 − 5

Name ______________________ Date ______________________

3 Subtraction Facts Minuends up to 10

Subtract.

1) 10 − 6	2) 9 − 2	3) 9 − 7	4) 10 − 1	5) 7 − 0
6) 3 − 3	7) 8 − 2	8) 10 − 3	9) 7 − 1	10) 10 − 5
11) 6 − 3	12) 7 − 5	13) 8 − 3	14) 9 − 6	15) 6 − 1
16) 9 − 3	17) 10 − 4	18) 10 − 7	19) 5 − 4	20) 6 − 2

SECTION 4

Fact Families Sums up to 10

12 worksheets

4 fact families per sheet

This section on fact families focuses only on the lower addition and subtraction facts with sums up to 10.

Name ______________________ Date ______________________

4 Fact Families Sums up to 10

Use the number sets to fill in the blanks and complete each fact family.

1)

2 + ___ = 5

3 + ___ = 5

___ - 2 = 3

5 - 3 = ___

2)

___ + 7 = 8

7 + 1 = ___

___ - 1 = 7

8 - ___ = 1

3)

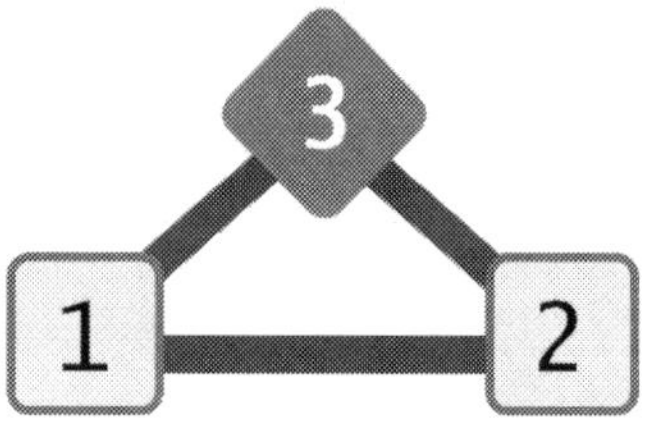

1 + 2 = ___

2 + ___ = 3

3 - 1 = ___

___ - 2 = 1

4)

5 + 5 = ___

___ - 5 = 5

Name ______________________ Date ______________________

4 Fact Families Sums up to 10

Use the number sets to fill in the blanks and complete each fact family.

1)

4 + 5 = ____

5 + ____ = 9

9 − 4 = ____

____ − 5 = 4

2)

____ + 8 = 10

8 + 2 = ____

10 − ____ = 8

____ − 8 = 2

3)

1 + 3 = ____

____ + 1 = 4

4 − ____ = 3

____ − 3 = 1

4)

____ + 4 = 7

4 + ____ = 7

____ − 3 = 4

7 − 4 = ____

Name ______________________ Date ______________________

4 Fact Families Sums up to 10

Use the number sets to fill in the blanks and complete each fact family.

1)

4 + 6 = ____

____ + 4 = 10

____ - 4 = 6

10 - ____ = 4

2)

____ + 4 = 5

4 + 1 = ____

____ - 1 = 4

5 - 4 = ____

3)

2 + 5 = ____

5 + ____ = 7

7 - 2 = ____

____ - 5 = 2

4)

2 + ____ = 8

6 + ____ = 8

____ - 2 = 6

8 - 6 = ____

Name ____________________ Date ____________________

4 Fact Families Sums up to 10

Use the number sets to fill in the blanks and complete each fact family.

1)

3 + 5 = ____

5 + ____ = 8

8 – 3 = ____

____ – 5 = 3

2)

____ + 7 = 9

7 + ____ = 9

____ – 2 = 7

9 – ____ = 2

3)

1 + ____ = 6

5 + ____ = 6

____ – 1 = 5

6 – ____ = 1

4)

2 + ____ = 4

____ – 2 = 2

Name ____________________ Date ____________________

4 Fact Families Sums up to 10

Use the number sets to fill in the blanks and complete each fact family.

1)

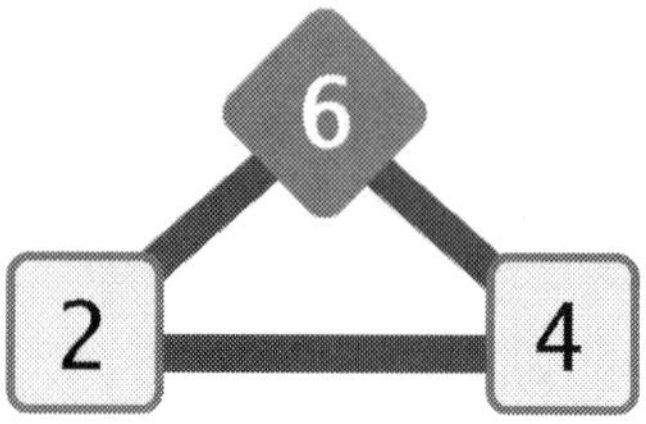

2 + ____ = 6

4 + ____ = 6

____ - 2 = 4

6 - ____ = 2

2)

____ + 4 = 8

8 - ____ = 4

3)

3 + 6 = ____

6 + ____ = 9

9 - 3 = ____

____ - 6 = 3

4)

1 + 9 = ____

____ + 1 = 10

10 - ____ = 9

10 - ____ = 1

Name ______________________ Date ______________________

4 Fact Families Sums up to 10

Use the number sets to fill in the blanks and complete each fact family.

1)

1 + 6 = ____

6 + ____ = 7

7 - 1 = ____

____ - 6 = 1

2)

____ + 7 = 10

7 + 3 = ____

10 - 3 = ____

10 - ____ = 3

3)

3 + ____ = 6

____ - 3 = 3

4)

____ + 8 = 9

8 + 1 = ____

____ - 1 = 8

9 - ____ = 1

Name ______________________ Date ______________________

4 Fact Families Sums up to 10

Use the number sets to fill in the blanks and write the fact families on your own.

1)

6

2 4

____

____

____

____

2)

8

3 5

____

____

____

____

3)

9

2 7

____

____

____

____

4)

10

4 6

____

____

____

____

Name ____________________ Date ____________________

4 Fact Families Sums up to 10

Use the number sets to fill in the blanks and write the fact families on your own.

1)

____

____

2)

____

____

____

____

3)

____

____

____

____

4)

____

____

____

____

Name ____________________ Date ____________________

4 Fact Families Sums up to 10

Use the number sets to fill in the blanks and write the fact families on your own.

1)

2)

3)

4)

Name ______________________ Date ______________________

4 Fact Families Sums up to 10

Use the number sets to fill in the blanks and write the fact families on your own.

1)

____

____

____

____

2)

____

____

3)

____

____

____

____

4)

____

____

____

____

Name ____________________ Date ____________________

4 Fact Families Sums up to 10

Use the number sets to fill in the blanks and write the fact families on your own.

1)

2)

3)

4)

Name ______________________ Date ______________________

4 Fact Families Sums up to 10

Use the number sets to fill in the blanks and write the fact families on your own.

____

____

____

____

____

____

____

____

____

____

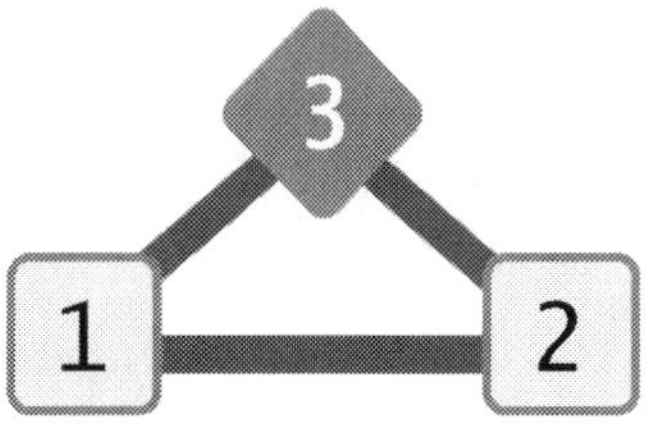

____

____

____

____

SECTION 5

Addition Facts Sums from 11 to 18

11 worksheets

20 problems per sheet

This section focuses only on the upper addition facts with sums from 11 to 18.

Name ______________________ Date ______________________

5 Addition Facts Sums from 11 to 18

Add.

1) $\begin{array}{r} 2 \\ +9 \\ \hline \end{array}$	2) $\begin{array}{r} 9 \\ +5 \\ \hline \end{array}$	3) $\begin{array}{r} 4 \\ +9 \\ \hline \end{array}$	4) $\begin{array}{r} 7 \\ +5 \\ \hline \end{array}$	5) $\begin{array}{r} 8 \\ +4 \\ \hline \end{array}$
6) $\begin{array}{r} 5 \\ +8 \\ \hline \end{array}$	7) $\begin{array}{r} 4 \\ +7 \\ \hline \end{array}$	8) $\begin{array}{r} 6 \\ +6 \\ \hline \end{array}$	9) $\begin{array}{r} 7 \\ +9 \\ \hline \end{array}$	10) $\begin{array}{r} 3 \\ +9 \\ \hline \end{array}$
11) $\begin{array}{r} 9 \\ +2 \\ \hline \end{array}$	12) $\begin{array}{r} 3 \\ +8 \\ \hline \end{array}$	13) $\begin{array}{r} 8 \\ +5 \\ \hline \end{array}$	14) $\begin{array}{r} 8 \\ +7 \\ \hline \end{array}$	15) $\begin{array}{r} 9 \\ +4 \\ \hline \end{array}$
16) $\begin{array}{r} 7 \\ +4 \\ \hline \end{array}$	17) $\begin{array}{r} 8 \\ +8 \\ \hline \end{array}$	18) $\begin{array}{r} 5 \\ +7 \\ \hline \end{array}$	19) $\begin{array}{r} 9 \\ +9 \\ \hline \end{array}$	20) $\begin{array}{r} 9 \\ +7 \\ \hline \end{array}$

Name ______________________ Date ______________________

5 Addition Facts Sums from 11 to 18

Add.

1) 9 + 4	2) 9 + 3	3) 5 + 9	4) 9 + 6	5) 4 + 9
6) 7 + 8	7) 2 + 9	8) 3 + 9	9) 6 + 5	10) 5 + 8
11) 6 + 6	12) 7 + 9	13) 6 + 8	14) 7 + 4	15) 7 + 6
16) 5 + 7	17) 7 + 5	18) 8 + 6	19) 9 + 8	20) 8 + 5

Name ____________________ Date ____________________

5 Addition Facts Sums from 11 to 18

Add.

1) 5 + 7	2) 9 + 7	3) 6 + 8	4) 5 + 6	5) 3 + 9
6) 7 + 4	7) 7 + 9	8) 8 + 8	9) 5 + 8	10) 6 + 5
11) 8 + 5	12) 9 + 6	13) 9 + 8	14) 2 + 9	15) 8 + 6
16) 6 + 6	17) 4 + 8	18) 6 + 9	19) 9 + 5	20) 4 + 9

Name ____________________ Date ____________________

5 Addition Facts Sums from 11 to 18

Add.

1) 6 + 6	2) 2 + 9	3) 3 + 9	4) 4 + 7	5) 7 + 8
6) 8 + 9	7) 4 + 9	8) 9 + 9	9) 9 + 2	10) 9 + 6
11) 7 + 4	12) 8 + 4	13) 9 + 5	14) 7 + 6	15) 8 + 5
16) 8 + 3	17) 4 + 8	18) 6 + 7	19) 3 + 8	20) 6 + 9

Name ______________________ Date ______________________

5 Addition Facts Sums from 11 to 18

Add.

1) 9 +6	2) 7 +8	3) 8 +7	4) 9 +7	5) 6 +8
6) 5 +6	7) 9 +8	8) 6 +6	9) 4 +8	10) 5 +9
11) 6 +7	12) 7 +9	13) 7 +4	14) 7 +6	15) 9 +2
16) 5 +7	17) 5 +8	18) 6 +9	19) 8 +8	20) 8 +5

Name ____________________ Date ____________________

5 Addition Facts Sums from 11 to 18

Add.

1) 5 + 6	2) 6 + 8	3) 9 + 3	4) 9 + 9	5) 2 + 9
6) 5 + 9	7) 6 + 6	8) 6 + 7	9) 4 + 7	10) 6 + 5
11) 4 + 8	12) 9 + 4	13) 8 + 5	14) 7 + 7	15) 7 + 5
16) 7 + 6	17) 7 + 8	18) 9 + 2	19) 5 + 7	20) 8 + 7

Name ______________________ Date ______________________

5 Addition Facts Sums from 11 to 18

Add.

1) $8 + 7$	2) $8 + 5$	3) $5 + 8$	4) $9 + 8$	5) $7 + 8$
6) $9 + 3$	7) $6 + 6$	8) $7 + 9$	9) $9 + 6$	10) $6 + 7$
11) $4 + 9$	12) $6 + 5$	13) $2 + 9$	14) $9 + 9$	15) $3 + 8$
16) $7 + 4$	17) $4 + 8$	18) $8 + 3$	19) $8 + 9$	20) $9 + 7$

Name ______________________ Date ______________________

5 Addition Facts Sums from 11 to 18

Add.

1) 8 + 3	2) 5 + 6	3) 9 + 5	4) 9 + 8	5) 7 + 8
6) 7 + 7	7) 8 + 5	8) 3 + 8	9) 9 + 7	10) 7 + 5
11) 4 + 8	12) 8 + 9	13) 7 + 6	14) 6 + 9	15) 3 + 9
16) 9 + 3	17) 9 + 4	18) 6 + 8	19) 7 + 9	20) 5 + 9

Name ____________________ Date ____________________

5 Addition Facts Sums from 11 to 18

Add.

1) 9 + 5	2) 9 + 9	3) 7 + 5	4) 5 + 8	5) 8 + 7
6) 8 + 6	7) 4 + 7	8) 6 + 7	9) 5 + 9	10) 9 + 7
11) 5 + 6	12) 8 + 4	13) 3 + 9	14) 6 + 6	15) 8 + 8
16) 5 + 7	17) 4 + 8	18) 2 + 9	19) 9 + 3	20) 4 + 9

Name ______________________________ Date ______________________________

5 Addition Facts Sums from 11 to 18

Add.

1) $\begin{array}{r} 4 \\ +8 \\ \hline \end{array}$	2) $\begin{array}{r} 8 \\ +9 \\ \hline \end{array}$	3) $\begin{array}{r} 6 \\ +5 \\ \hline \end{array}$	4) $\begin{array}{r} 8 \\ +3 \\ \hline \end{array}$	5) $\begin{array}{r} 8 \\ +5 \\ \hline \end{array}$
6) $\begin{array}{r} 6 \\ +8 \\ \hline \end{array}$	7) $\begin{array}{r} 6 \\ +7 \\ \hline \end{array}$	8) $\begin{array}{r} 9 \\ +7 \\ \hline \end{array}$	9) $\begin{array}{r} 7 \\ +9 \\ \hline \end{array}$	10) $\begin{array}{r} 8 \\ +7 \\ \hline \end{array}$
11) $\begin{array}{r} 2 \\ +9 \\ \hline \end{array}$	12) $\begin{array}{r} 9 \\ +6 \\ \hline \end{array}$	13) $\begin{array}{r} 5 \\ +6 \\ \hline \end{array}$	14) $\begin{array}{r} 7 \\ +8 \\ \hline \end{array}$	15) $\begin{array}{r} 8 \\ +6 \\ \hline \end{array}$
16) $\begin{array}{r} 7 \\ +7 \\ \hline \end{array}$	17) $\begin{array}{r} 5 \\ +9 \\ \hline \end{array}$	18) $\begin{array}{r} 7 \\ +5 \\ \hline \end{array}$	19) $\begin{array}{r} 8 \\ +4 \\ \hline \end{array}$	20) $\begin{array}{r} 3 \\ +9 \\ \hline \end{array}$

Name ______________________ Date ______________________

5 Addition Facts Sums from 11 to 18

Add.

1) $\begin{array}{r} 7 \\ +9 \\ \hline \end{array}$	2) $\begin{array}{r} 8 \\ +8 \\ \hline \end{array}$	3) $\begin{array}{r} 2 \\ +9 \\ \hline \end{array}$	4) $\begin{array}{r} 8 \\ +4 \\ \hline \end{array}$	5) $\begin{array}{r} 9 \\ +8 \\ \hline \end{array}$
6) $\begin{array}{r} 9 \\ +9 \\ \hline \end{array}$	7) $\begin{array}{r} 4 \\ +8 \\ \hline \end{array}$	8) $\begin{array}{r} 9 \\ +7 \\ \hline \end{array}$	9) $\begin{array}{r} 7 \\ +4 \\ \hline \end{array}$	10) $\begin{array}{r} 8 \\ +5 \\ \hline \end{array}$
11) $\begin{array}{r} 5 \\ +6 \\ \hline \end{array}$	12) $\begin{array}{r} 7 \\ +8 \\ \hline \end{array}$	13) $\begin{array}{r} 6 \\ +6 \\ \hline \end{array}$	14) $\begin{array}{r} 6 \\ +9 \\ \hline \end{array}$	15) $\begin{array}{r} 6 \\ +8 \\ \hline \end{array}$
16) $\begin{array}{r} 5 \\ +8 \\ \hline \end{array}$	17) $\begin{array}{r} 3 \\ +8 \\ \hline \end{array}$	18) $\begin{array}{r} 4 \\ +7 \\ \hline \end{array}$	19) $\begin{array}{r} 7 \\ +5 \\ \hline \end{array}$	20) $\begin{array}{r} 6 \\ +7 \\ \hline \end{array}$

SECTION 6

Missing Addends Sums from 11 to 18

11 worksheets

15 problems per sheet

This fill-in-the-blank section helps with the transition from addition to subtraction and focuses only on the upper addition facts with sums from 11 to 18.

Name ______________________ Date ______________________

6 Missing Addends Sums from 11 to 18

Fill in the blanks with the missing addends.

1) 5+____=11 2) 6+____=11 3) 6+____=14

4) 7+____=11 5) 8+____=16 6) 4+____=13

7) 5+____=14 8) 8+____=13 9) 5+____=13

10) 8+____=15 11) 8+____=12 12) 3+____=11

13) 6+____=15 14) 9+____=18 15) 9+____=17

Name ______________________ Date ______________________

6 Missing Addends Sums from 11 to 18

Fill in the blanks with the missing addends.

1) 9+____=11 2) 9+____=12 3) 5+____=13

4) 6+____=15 5) 7+____=13 6) 7+____=12

7) 8+____=13 8) 9+____=14 9) 9+____=16

10) 2+____=11 11) 6+____=14 12) 9+____=13

13) 7+____=14 14) 5+____=12 15) 6+____=11

Name ______________________ Date ______________________

6 Missing Addends Sums from 11 to 18

Fill in the blanks with the missing addends.

1) 2+____=11 2) 7+____=15 3) 8+____=12

4) 8+____=13 5) 6+____=11 6) 7+____=14

7) 3+____=12 8) 6+____=15 9) 5+____=14

10) 5+____=12 11) 9+____=11 12) 9+____=15

13) 8+____=16 14) 8+____=15 15) 6+____=14

Name ________________ Date ________________

6 Missing Addends Sums from 11 to 18

Fill in the blanks with the missing addends.

1) 8+____=15 2) 9+____=14 3) 7+____=15

4) 3+____=11 5) 8+____=16 6) 5+____=13

7) 2+____=11 8) 8+____=17 9) 3+____=12

10) 9+____=16 11) 4+____=11 12) 7+____=11

13) 6+____=14 14) 9+____=12 15) 6+____=15

Name ____________________ Date ____________________

6 Missing Addends Sums from 11 to 18

Fill in the blanks with the missing addends.

1) 9+____=16 2) 5+____=13 3) 8+____=13

4) 8+____=15 5) 7+____=11 6) 7+____=13

7) 8+____=17 8) 7+____=14 9) 5+____=12

10) 9+____=13 11) 4+____=13 12) 3+____=12

13) 9+____=18 14) 7+____=15 15) 7+____=16

Name ____________________ Date ____________________

6 Missing Addends Sums from 11 to 18

Fill in the blanks with the missing addends.

1) 5+____=12 2) 2+____=11 3) 4+____=11

4) 8+____=16 5) 8+____=15 6) 6+____=15

7) 8+____=13 8) 7+____=13 9) 9+____=12

10) 8+____=14 11) 8+____=12 12) 4+____=12

13) 6+____=12 14) 9+____=17 15) 3+____=12

Name ______________________ Date ______________________

6 Missing Addends Sums from 11 to 18

Fill in the blanks with the missing addends.

1) $7 + \_\_\_\_ = 16$ 2) $6 + \_\_\_\_ = 13$ 3) $5 + \_\_\_\_ = 11$

4) $4 + \_\_\_\_ = 13$ 5) $3 + \_\_\_\_ = 12$ 6) $9 + \_\_\_\_ = 14$

7) $9 + \_\_\_\_ = 13$ 8) $5 + \_\_\_\_ = 14$ 9) $8 + \_\_\_\_ = 15$

10) $9 + \_\_\_\_ = 15$ 11) $7 + \_\_\_\_ = 14$ 12) $5 + \_\_\_\_ = 12$

13) $5 + \_\_\_\_ = 13$ 14) $9 + \_\_\_\_ = 11$ 15) $9 + \_\_\_\_ = 17$

Name ____________________ Date ____________________

6 Missing Addends Sums from 11 to 18

Fill in the blanks with the missing addends.

1) 9+____=15 2) 7+____=11 3) 4+____=12

4) 9+____=14 5) 3+____=11 6) 6+____=11

7) 6+____=13 8) 9+____=16 9) 6+____=15

10) 9+____=12 11) 7+____=16 12) 7+____=15

13) 8+____=15 14) 7+____=13 15) 5+____=12

Name ______________________ Date ______________________

6 Missing Addends Sums from 11 to 18

Fill in the blanks with the missing addends.

1) 4+____=12 2) 4+____=11 3) 9+____=11

4) 8+____=15 5) 6+____=13 6) 5+____=11

7) 5+____=12 8) 8+____=11 9) 3+____=12

10) 4+____=13 11) 8+____=12 12) 9+____=16

13) 6+____=12 14) 3+____=11 15) 8+____=17

Name ______________________ Date ______________________

6 Missing Addends Sums from 11 to 18

Fill in the blanks with the missing addends.

1) 8+____=14 2) 8+____=11 3) 9+____=12

4) 5+____=12 5) 2+____=11 6) 5+____=13

7) 7+____=15 8) 3+____=11 9) 9+____=18

10) 8+____=12 11) 6+____=13 12) 6+____=15

13) 8+____=17 14) 9+____=15 15) 9+____=16

Name ______________________ Date ______________________

6 Missing Addends Sums from 11 to 18

Fill in the blanks with the missing addends.

1) 8+____=14 2) 7+____=14 3) 8+____=12

4) 7+____=11 5) 9+____=13 6) 3+____=11

7) 8+____=13 8) 9+____=11 9) 9+____=14

10) 6+____=13 11) 4+____=12 12) 8+____=16

13) 9+____=18 14) 8+____=11 15) 6+____=15

SECTION 7

Subtraction Facts Minuends from 11 to 18

11 worksheets

20 problems per sheet

This section focuses only on the upper subtraction facts with minuends from 11 to 18.

Name ______________________ Date ______________________

7 Subtraction Facts Minuends from 11 to 18

Subtract.

1) 11 −7	2) 15 −8	3) 17 −8	4) 11 −5	5) 11 −4
6) 16 −9	7) 15 −7	8) 14 −8	9) 11 −9	10) 13 −4
11) 17 −9	12) 12 −6	13) 14 −7	14) 12 −5	15) 12 −4
16) 12 −9	17) 14 −9	18) 11 −8	19) 16 −8	20) 12 −3

Name ____________________ Date ____________________

7 Subtraction Facts Minuends from 11 to 18

Subtract.

1) 12 − 9

2) 12 − 5

3) 14 − 9

4) 11 − 3

5) 14 − 8

6) 11 − 9

7) 11 − 6

8) 17 − 9

9) 12 − 7

10) 15 − 8

11) 13 − 9

12) 11 − 5

13) 17 − 8

14) 12 − 6

15) 13 − 5

16) 18 − 9

17) 12 − 8

18) 15 − 9

19) 14 − 5

20) 11 − 8

Name ______________________ Date ______________________

7 Subtraction Facts Minuends from 11 to 18

Subtract.

1) 15 −6	2) 12 −7	3) 15 −7	4) 17 −8	5) 12 −8
6) 12 −4	7) 12 −5	8) 14 −6	9) 18 −9	10) 17 −9
11) 16 −7	12) 11 −8	13) 11 −6	14) 13 −8	15) 13 −7
16) 11 −4	17) 13 −6	18) 14 −7	19) 14 −8	20) 12 −6

Name ______________________ Date ______________________

7 Subtraction Facts Minuends from 11 to 18

Subtract.

1) 14 − 9	2) 11 − 4	3) 16 − 7	4) 16 − 8	5) 11 − 3
6) 11 − 6	7) 13 − 8	8) 12 − 4	9) 13 − 5	10) 11 − 2
11) 14 − 7	12) 18 − 9	13) 11 − 5	14) 12 − 8	15) 11 − 9
16) 15 − 9	17) 12 − 7	18) 17 − 9	19) 12 − 5	20) 14 − 5

Name ______________________ Date ______________________

7 Subtraction Facts Minuends from 11 to 18

Subtract.

1) 13 − 7	2) 12 − 3	3) 13 − 8	4) 15 − 7	5) 12 − 5
6) 14 − 7	7) 14 − 8	8) 15 − 6	9) 11 − 9	10) 12 − 8
11) 14 − 6	12) 13 − 4	13) 11 − 5	14) 11 − 8	15) 18 − 9
16) 11 − 3	17) 15 − 8	18) 11 − 4	19) 16 − 7	20) 16 − 8

Name ______________________ Date ______________________

7 Subtraction Facts Minuends from 11 to 18

Subtract.

1) 16 − 9	2) 11 − 3	3) 14 − 7	4) 16 − 8	5) 12 − 6
6) 18 − 9	7) 16 − 7	8) 14 − 6	9) 12 − 4	10) 13 − 4
11) 13 − 9	12) 13 − 7	13) 14 − 5	14) 14 − 9	15) 11 − 9
16) 12 − 9	17) 13 − 6	18) 15 − 7	19) 14 − 8	20) 11 − 6

Name ______________________ Date ______________________

7 Subtraction Facts Minuends from 11 to 18

Subtract.

1) 11 −9	2) 13 −4	3) 14 −5	4) 16 −9	5) 15 −7
6) 14 −7	7) 12 −6	8) 13 −9	9) 15 −8	10) 18 −9
11) 12 −9	12) 12 −3	13) 11 −8	14) 11 −2	15) 14 −9
16) 12 −8	17) 17 −9	18) 12 −5	19) 11 −3	20) 14 −8

Name ______________________ Date ______________________

7 Subtraction Facts Minuends from 11 to 18

Subtract.

1) 15 − 7	2) 12 − 9	3) 11 − 2	4) 13 − 4	5) 12 − 6
6) 12 − 5	7) 14 − 7	8) 11 − 3	9) 15 − 9	10) 12 − 3
11) 13 − 9	12) 11 − 4	13) 16 − 8	14) 11 − 8	15) 12 − 4
16) 17 − 9	17) 12 − 8	18) 17 − 8	19) 14 − 9	20) 13 − 7

Name ______________________ Date ______________________

7 Subtraction Facts Minuends from 11 to 18

Subtract.

1) 12 − 6	2) 11 − 9	3) 12 − 8	4) 15 − 6	5) 11 − 2
6) 12 − 7	7) 11 − 5	8) 11 − 7	9) 12 − 9	10) 13 − 6
11) 12 − 3	12) 12 − 4	13) 14 − 7	14) 14 − 5	15) 15 − 7
16) 17 − 9	17) 13 − 4	18) 11 − 6	19) 14 − 9	20) 11 − 4

Name ______________________ Date ______________________

7 Subtraction Facts Minuends from 11 to 18

Subtract.

1) 13 − 8	2) 11 − 3	3) 12 − 7	4) 17 − 9	5) 15 − 9
6) 13 − 6	7) 11 − 9	8) 14 − 8	9) 12 − 4	10) 15 − 7
11) 18 − 9	12) 11 − 6	13) 16 − 9	14) 11 − 7	15) 15 − 6
16) 17 − 8	17) 12 − 5	18) 14 − 7	19) 11 − 4	20) 12 − 9

Name ______________________ Date ______________________

7 Subtraction Facts Minuends from 11 to 18

Subtract.

1) 12 −4	2) 11 −6	3) 17 −9	4) 14 −7	5) 11 −4
6) 15 −6	7) 14 −9	8) 16 −7	9) 11 −8	10) 13 −4
11) 16 −8	12) 14 −5	13) 11 −9	14) 15 −8	15) 18 −9
16) 12 −7	17) 14 −8	18) 13 −8	19) 12 −9	20) 15 −7

SECTION 8

Fact Families Sums from 11 to 18

10 worksheets

4 fact families per sheet

This section on fact families focuses only on the upper addition and subtraction facts with sums from 11 to 18.

Name ______________________ Date ______________________

8 Fact Families Sums from 11 to 18

Use the number sets to fill in the blanks and complete each fact family.

1)

3 + 9 = ____

____ + 3 = 12

____ − 3 = 9

12 − ____ = 3

2)

____ + 9 = 17

9 + 8 = ____

17 − ____ = 9

____ − 9 = 8

3)

6 + ____ = 13

7 + 6 = ____

____ − 6 = 7

13 − 7 = ____

4)

5 + 6 = ____

____ + 5 = 11

11 − ____ = 6

____ − 6 = 5

Name ______________________ Date ______________________

8 Fact Families Sums from 11 to 18

Use the number sets to fill in the blanks and complete each fact family.

1)

____ + 9 = 14

9 + 5 = ____

14 - ____ = 9

____ - 9 = 5

2)

7 + ____ = 15

8 + 7 = ____

____ - 7 = 8

15 - 8 = ____

3)

2 + 9 = ____

____ + 2 = 11

11 - ____ = 9

____ - 9 = 2

4)

8 + ____ = 16

____ - 8 = 8

Name ____________________ Date ____________________

8 Fact Families Sums from 11 to 18

Use the number sets to fill in the blanks and complete each fact family.

1)

6 + 8 = ____

____ + 6 = 14

____ - 6 = 8

14 - 8 = ____

2)

____ + 8 = 12

8 + ____ = 12

12 - 4 = ____

____ - 8 = 4

3)

____ + 7 = 11

7 + ____ = 11

____ - 4 = 7

11 - ____ = 4

4)

____ + 9 = 18

18 - 9 = ____

Name ____________________ Date ____________________

8 Fact Families Sums from 11 to 18

Use the number sets to fill in the blanks and complete each fact family.

1)

6 + 6 = ____

____ - 6 = 6

2)

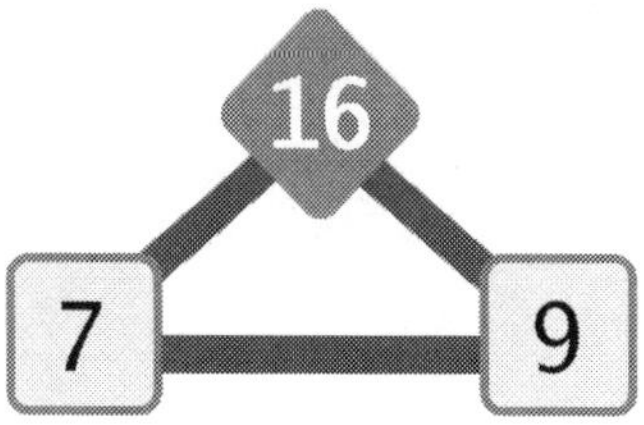

7 + 9 = ____

____ + 7 = 16

____ - 7 = 9

16 - ____ = 7

3)

____ + 9 = 13

9 + 4 = ____

13 - ____ = 9

____ - 9 = 4

4)

3 + 8 = ____

____ + 3 = 11

____ - 3 = 8

11 - ____ = 3

Name ______________________ Date ______________________

8 Fact Families Sums from 11 - 18

Use the number sets to fill in the blanks and complete each fact family.

1)

6 + ____ = 15

9 + 6 = ____

____ - 6 = 9

15 - ____ = 6

2)

7 + ____ = 14

14 - 7 = ____

3)

5 + ____ = 13

____ + 5 = 13

13 - ____ = 8

____ - 8 = 5

4)

5 + 7 = ____

____ + 5 = 12

12 - ____ = 7

____ - 7 = 5

Name ______________________ Date ______________________

8 Fact Families Sums from 11 to 18

Use the number sets to fill in the blanks and write the fact families on your own.

1)

2)

3)

4)

Name ____________________ Date ____________________

8 Fact Families Sums from 11 to 18

Use the number sets to fill in the blanks and write the fact families on your own.

1)

____

____

____

____

2)

____

____

____

____

3)

____

____

4)

____

____

____

____

Name ______________ Date ______________

8 Fact Families Sums from 11 to 18

Use the number sets to fill in the blanks and write the fact families on your own.

1)

2)

3)

4)

Name ____________________ Date ____________________

8 Fact Families Sums from 11 to 18

Use the number sets to fill in the blanks and write the fact families on your own.

1)

____

____

____

____

2)

____

____

____

____

3)

____

____

4)

____

____

____

____

Name ______________________ Date ______________________

8 Fact Families Sums from 11 to 18

Use the number sets to fill in the blanks and write the fact families on your own.

1)

____

____

____

____

2)

____

____

3)

____

____

____

____

4)

____

____

____

____

SECTION 9

Addition Facts All Sums up to 18

11 worksheets

20 problems per sheet

This section combines the lower and upper addition facts with sums up to 18.

Name ________________ Date ________________

9 Addition Facts Sums from 0-18

Add.

1) 1 +3	2) 5 +4	3) 3 +8	4) 9 +1	5) 7 +5
6) 4 +7	7) 7 +3	8) 1 +5	9) 1 +9	10) 7 +6
11) 7 +7	12) 9 +7	13) 9 +6	14) 7 +8	15) 8 +0
16) 3 +5	17) 8 +9	18) 5 +3	19) 6 +9	20) 6 +8

Name ____________________ Date ____________________

9 Addition Facts Sums from 0-18

Add.

1) 2 + 0	2) 6 + 9	3) 7 + 8	4) 4 + 1	5) 9 + 2
6) 3 + 7	7) 2 + 9	8) 1 + 5	9) 4 + 7	10) 3 + 6
11) 4 + 5	12) 6 + 4	13) 8 + 5	14) 0 + 4	15) 9 + 3
16) 7 + 7	17) 0 + 7	18) 6 + 8	19) 6 + 2	20) 2 + 3

Name ______________________ Date ______________________

9 Addition Facts Sums from 0-18

Add.

1) 7 +1	2) 8 +5	3) 2 +8	4) 7 +3	5) 8 +2
6) 2 +7	7) 9 +7	8) 6 +7	9) 2 +4	10) 6 +5
11) 8 +7	12) 2 +1	13) 9 +9	14) 6 +9	15) 5 +8
16) 1 +9	17) 9 +6	18) 2 +2	19) 3 +9	20) 3 +3

Name ______________________ Date ______________________

9 Addition Facts Sums from 0-18

Add.

1) $\begin{array}{r} 9 \\ +9 \\ \hline \end{array}$	2) $\begin{array}{r} 7 \\ +9 \\ \hline \end{array}$	3) $\begin{array}{r} 0 \\ +3 \\ \hline \end{array}$	4) $\begin{array}{r} 8 \\ +9 \\ \hline \end{array}$	5) $\begin{array}{r} 6 \\ +7 \\ \hline \end{array}$
6) $\begin{array}{r} 2 \\ +5 \\ \hline \end{array}$	7) $\begin{array}{r} 3 \\ +4 \\ \hline \end{array}$	8) $\begin{array}{r} 1 \\ +3 \\ \hline \end{array}$	9) $\begin{array}{r} 7 \\ +4 \\ \hline \end{array}$	10) $\begin{array}{r} 5 \\ +1 \\ \hline \end{array}$
11) $\begin{array}{r} 9 \\ +5 \\ \hline \end{array}$	12) $\begin{array}{r} 7 \\ +1 \\ \hline \end{array}$	13) $\begin{array}{r} 5 \\ +6 \\ \hline \end{array}$	14) $\begin{array}{r} 9 \\ +8 \\ \hline \end{array}$	15) $\begin{array}{r} 4 \\ +3 \\ \hline \end{array}$
16) $\begin{array}{r} 8 \\ +5 \\ \hline \end{array}$	17) $\begin{array}{r} 6 \\ +3 \\ \hline \end{array}$	18) $\begin{array}{r} 8 \\ +4 \\ \hline \end{array}$	19) $\begin{array}{r} 2 \\ +2 \\ \hline \end{array}$	20) $\begin{array}{r} 5 \\ +4 \\ \hline \end{array}$

Name ____________________ Date ____________________

9 Addition Facts Sums from 0-18

Add.

1) 4 + 3	2) 7 + 8	3) 5 + 5	4) 5 + 8	5) 6 + 6
6) 2 + 9	7) 8 + 7	8) 0 + 8	9) 5 + 1	10) 7 + 9
11) 1 + 1	12) 6 + 0	13) 7 + 3	14) 6 + 1	15) 9 + 8
16) 5 + 4	17) 5 + 9	18) 9 + 6	19) 9 + 1	20) 9 + 9

Name ______________________ Date ______________________

9 Addition Facts Sums from 0-18

Add.

1) 1 + 5	2) 6 + 3	3) 7 + 5	4) 0 + 3	5) 3 + 2
6) 5 + 0	7) 6 + 8	8) 4 + 9	9) 3 + 6	10) 3 + 5
11) 9 + 9	12) 7 + 8	13) 4 + 1	14) 2 + 9	15) 6 + 6
16) 4 + 7	17) 6 + 7	18) 2 + 8	19) 1 + 9	20) 8 + 8

Name ______________________ Date ______________________

9 Addition Facts Sums from 0-18

Add.

1) 5 +8	2) 9 +8	3) 3 +7	4) 2 +4	5) 8 +0
6) 3 +9	7) 6 +6	8) 5 +3	9) 7 +5	10) 9 +1
11) 7 +3	12) 1 +9	13) 4 +1	14) 7 +9	15) 3 +4
16) 2 +7	17) 9 +3	18) 5 +7	19) 8 +5	20) 8 +8

Name ______________________ Date ______________________

9 Addition Facts Sums from 0-18

Add.

1) 7 + 9	2) 3 + 6	3) 3 + 5	4) 5 + 2	5) 6 + 8
6) 2 + 5	7) 5 + 8	8) 7 + 2	9) 4 + 8	10) 7 + 5
11) 5 + 6	12) 1 + 0	13) 1 + 2	14) 9 + 4	15) 9 + 6
16) 8 + 6	17) 6 + 5	18) 6 + 6	19) 4 + 5	20) 2 + 3

Name ______________________ Date ______________________

9 Addition Facts Sums from 0-18

Add.

1) 1 +4	2) 5 +6	3) 6 +3	4) 9 +2	5) 8 +5
6) 9 +6	7) 0 +2	8) 2 +8	9) 1 +2	10) 3 +7
11) 7 +6	12) 2 +9	13) 4 +4	14) 9 +4	15) 2 +1
16) 7 +2	17) 7 +8	18) 8 +2	19) 7 +7	20) 6 +9

Name ______________________ Date ______________________

9 Addition Facts Sums from 0-18

Add.

1) 2 + 5 = ____

2) 5 + 1 = ____

3) 7 + 7 = ____

4) 7 + 6 = ____

5) 2 + 4 = ____

6) 2 + 3 = ____

7) 0 + 0 = ____

8) 9 + 9 = ____

9) 5 + 6 = ____

10) 4 + 9 = ____

11) 3 + 9 = ____

12) 1 + 4 = ____

13) 1 + 7 = ____

14) 4 + 6 = ____

15) 8 + 6 = ____

16) 5 + 8 = ____

17) 3 + 6 = ____

18) 6 + 7 = ____

19) 7 + 9 = ____

20) 6 + 9 = ____

Name ____________________ Date ____________________

9 Addition Facts Sums from 0-18

Add.

1) 3 + 8	2) 8 + 4	3) 1 + 5	4) 9 + 7	5) 3 + 9
6) 2 + 5	7) 4 + 6	8) 5 + 3	9) 0 + 4	10) 1 + 7
11) 6 + 4	12) 4 + 8	13) 5 + 2	14) 9 + 4	15) 7 + 6
16) 7 + 4	17) 9 + 2	18) 2 + 3	19) 2 + 7	20) 3 + 2

SECTION 10

Missing Addends All Sums up to 18

11 worksheets

15 problems per sheet

This fill-in-the-blank section helps with the transition from addition to subtraction and is a combination of lower and upper addition facts with sums from 11 to 18.

Name ______________________ Date ______________________

10 Missing Addends Sums from 0 to 18

Fill in the blanks with the missing addends.

1) $5 + \_\_\_\_ = 10$ 2) $3 + \_\_\_\_ = 7$ 3) $2 + \_\_\_\_ = 6$

4) $8 + \_\_\_\_ = 15$ 5) $9 + \_\_\_\_ = 17$ 6) $9 + \_\_\_\_ = 13$

7) $3 + \_\_\_\_ = 6$ 8) $5 + \_\_\_\_ = 13$ 9) $1 + \_\_\_\_ = 7$

10) $8 + \_\_\_\_ = 10$ 11) $2 + \_\_\_\_ = 11$ 12) $8 + \_\_\_\_ = 11$

13) $3 + \_\_\_\_ = 5$ 14) $9 + \_\_\_\_ = 15$ 15) $7 + \_\_\_\_ = 15$

Name ______________________ Date ______________________

10 Missing Addends Sums from 0 to 18

Fill in the blanks with the missing addends.

1) 3+____=11 2) 4+____=8 3) 7+____=8

4) 0+____=6 5) 7+____=12 6) 3+____=12

7) 4+____=12 8) 6+____=15 9) 8+____=16

10) 5+____=9 11) 8+____=17 12) 8+____=10

13) 2+____=11 14) 2+____=5 15) 6+____=14

Name ______________________ Date ______________________

10 Missing Addends Sums from 0 to 18

Fill in the blanks with the missing addends.

1) 2+____=8 2) 2+____=4 3) 6+____=7

4) 8+____=14 5) 9+____=13 6) 7+____=11

7) 0+____=3 8) 5+____=12 9) 3+____=5

10) 6+____=10 11) 8+____=17 12) 6+____=11

13) 4+____=7 14) 2+____=7 15) 1+____=10

Name ______________________ Date ______________________

10 Missing Addends Sums from 0 to 18

Fill in the blanks with the missing addends.

1) 5+____=13 2) 4+____=5 3) 6+____=10

4) 3+____=4 5) 9+____=18 6) 0+____=4

7) 6+____=9 8) 7+____=12 9) 8+____=15

10) 9+____=17 11) 5+____=14 12) 7+____=16

13) 6+____=12 14) 2+____=10 15) 6+____=8

Name ______________________ Date ______________________

10 Missing Addends Sums from 0 to 18

Fill in the blanks with the missing addends.

1) 6+____=14 2) 4+____=12 3) 4+____=6

4) 3+____=4 5) 5+____=9 6) 2+____=7

7) 5+____=6 8) 8+____=16 9) 2+____=3

10) 6+____=13 11) 9+____=11 12) 8+____=11

13) 6+____=10 14) 4+____=10 15) 9+____=16

Name ______________________ Date ______________________

10 Missing Addends Sums from 0 to 18

Fill in the blanks with the missing addends.

1) 8+____=11 2) 4+____=7 3) 8+____=10

4) 8+____=15 5) 5+____=9 6) 0+____=4

7) 1+____=8 8) 3+____=12 9) 5+____=8

10) 3+____=8 11) 4+____=10 12) 9+____=11

13) 2+____=8 14) 7+____=11 15) 1+____=9

Name ____________________ Date ____________________

10 Missing Addends Sums from 0 to 18

Fill in the blanks with the missing addends.

1) $7+\_\_\_\_=7$ 2) $3+\_\_\_\_=10$ 3) $7+\_\_\_\_=12$

4) $4+\_\_\_\_=9$ 5) $4+\_\_\_\_=10$ 6) $8+\_\_\_\_=12$

7) $5+\_\_\_\_=8$ 8) $1+\_\_\_\_=4$ 9) $3+\_\_\_\_=5$

10) $3+\_\_\_\_=12$ 11) $1+\_\_\_\_=2$ 12) $3+\_\_\_\_=6$

13) $9+\_\_\_\_=17$ 14) $6+\_\_\_\_=10$ 15) $8+\_\_\_\_=14$

Name ______________________________ Date ______________________________

10 Missing Addends Sums from 0 to 18

Fill in the blanks with the missing addends.

1) 1+____=8 2) 2+____=3 3) 6+____=14

4) 9+____=14 5) 6+____=8 6) 3+____=9

7) 9+____=12 8) 2+____=6 9) 6+____=10

10) 8+____=9 11) 8+____=8 12) 4+____=12

13) 6+____=9 14) 2+____=8 15) 5+____=11

Name ______________________ Date ______________________

10 Missing Addends Sums from 0 to 18

Fill in the blanks with the missing addends.

1) 9+____=17 2) 6+____=8 3) 1+____=5

4) 1+____=3 5) 5+____=8 6) 1+____=2

7) 3+____=10 8) 2+____=4 9) 9+____=15

10) 0+____=5 11) 4+____=11 12) 6+____=9

13) 8+____=15 14) 6+____=11 15) 2+____=11

Name ______________________ Date ______________________

10 Missing Addends Sums from 0 to 18

Fill in the blanks with the missing addends.

1) 3+____=11   2) 2+____=6   3) 0+____=5

4) 4+____=9   5) 9+____=17   6) 3+____=6

7) 8+____=15   8) 8+____=12   9) 4+____=11

10) 1+____=8   11) 9+____=16   12) 2+____=11

13) 6+____=15   14) 5+____=10   15) 3+____=12

Name ______________________ Date ______________________

10 Missing Addends Sums from 0 to 18

Fill in the blanks with the missing addends.

1) $1+\_\_\_\_=7$ 2) $3+\_\_\_\_=5$ 3) $1+\_\_\_\_=4$

4) $0+\_\_\_\_=1$ 5) $4+\_\_\_\_=10$ 6) $9+\_\_\_\_=16$

7) $8+\_\_\_\_=16$ 8) $6+\_\_\_\_=15$ 9) $5+\_\_\_\_=8$

10) $3+\_\_\_\_=8$ 11) $4+\_\_\_\_=12$ 12) $7+\_\_\_\_=9$

13) $3+\_\_\_\_=6$ 14) $3+\_\_\_\_=11$ 15) $8+\_\_\_\_=12$

SECTION 11

Subtraction Facts Minuends up to 18

11 worksheets

20 problems per sheet

This section combines the lower and upper subtraction facts with minuends up to 18.

Name ________________ Date ________________

11 Subtraction Facts Minuends up to 18

Subtract.

1) 15 − 6	2) 11 − 3	3) 10 − 1	4) 9 − 2	5) 6 − 2
6) 12 − 7	7) 14 − 5	8) 16 − 9	9) 8 − 5	10) 7 − 1
11) 15 − 7	12) 5 − 5	13) 3 − 0	14) 14 − 9	15) 13 − 8
16) 4 − 2	17) 8 − 0	18) 6 − 1	19) 6 − 6	20) 13 − 5

Name ______________________ Date ______________________

11 Subtraction Facts Minuends up to 18

Subtract.

1) $6 - 3$	2) $12 - 5$	3) $12 - 3$	4) $10 - 3$	5) $2 - 1$
6) $3 - 0$	7) $10 - 2$	8) $10 - 1$	9) $5 - 4$	10) $4 - 1$
11) $13 - 4$	12) $7 - 0$	13) $14 - 8$	14) $5 - 5$	15) $12 - 6$
16) $0 - 0$	17) $9 - 6$	18) $13 - 7$	19) $8 - 2$	20) $14 - 5$

Name ______________________ Date ______________________

11 Subtraction Facts Minuends up to 18

Subtract.

1) 14 − 9	2) 11 − 7	3) 8 − 7	4) 10 − 9	5) 11 − 4
6) 12 − 8	7) 8 − 2	8) 5 − 4	9) 12 − 4	10) 11 − 8
11) 7 − 7	12) 2 − 0	13) 15 − 6	14) 2 − 2	15) 18 − 9
16) 16 − 9	17) 6 − 0	18) 1 − 0	19) 12 − 6	20) 10 − 2

Name ______________________ Date ______________________

11 Subtraction Facts Minuends up to 18

Subtract.

1) 10 − 3	2) 13 − 4	3) 6 − 3	4) 4 − 2	5) 9 − 6
6) 9 − 2	7) 10 − 9	8) 15 − 8	9) 16 − 8	10) 6 − 4
11) 13 − 7	12) 12 − 9	13) 5 − 3	14) 14 − 8	15) 2 − 1
16) 1 − 0	17) 15 − 9	18) 6 − 5	19) 5 − 2	20) 7 − 6

Name ______________________ Date ______________________

11 Subtraction Facts Minuends up to 18

Subtract.

1) 10 −5	2) 8 −6	3) 4 −2	4) 6 −1	5) 11 −7
6) 15 −7	7) 14 −8	8) 3 −1	9) 7 −3	10) 7 −5
11) 17 −9	12) 13 −8	13) 10 −7	14) 8 −3	15) 12 −3
16) 2 −1	17) 14 −5	18) 6 −0	19) 11 −6	20) 11 −9

Name ______________________ Date ______________________

11 Subtraction Facts Minuends up to 18

Subtract.

1) 7 − 5	2) 15 − 8	3) 9 − 0	4) 14 − 8	5) 4 − 3
6) 10 − 4	7) 16 − 7	8) 6 − 1	9) 14 − 9	10) 13 − 5
11) 6 − 3	12) 7 − 3	13) 5 − 5	14) 16 − 9	15) 4 − 1
16) 3 − 0	17) 10 − 8	18) 15 − 9	19) 17 − 9	20) 9 − 8

Name ____________ Date ____________

11 Subtraction Facts Minuends up to 18

Subtract.

1) 15 − 6	2) 1 − 1	3) 10 − 6	4) 9 − 7	5) 17 − 8
6) 2 − 0	7) 12 − 5	8) 7 − 0	9) 6 − 1	10) 12 − 9
11) 14 − 9	12) 9 − 6	13) 6 − 0	14) 4 − 3	15) 17 − 9
16) 15 − 8	17) 11 − 3	18) 13 − 8	19) 6 − 4	20) 14 − 6

Name ______________________ Date ______________________

11 Subtraction Facts Minuends up to 18

Subtract.

1) $\begin{array}{r} 8 \\ -4 \\ \hline \end{array}$	2) $\begin{array}{r} 13 \\ -5 \\ \hline \end{array}$	3) $\begin{array}{r} 14 \\ -6 \\ \hline \end{array}$	4) $\begin{array}{r} 7 \\ -5 \\ \hline \end{array}$	5) $\begin{array}{r} 2 \\ -0 \\ \hline \end{array}$
6) $\begin{array}{r} 11 \\ -6 \\ \hline \end{array}$	7) $\begin{array}{r} 12 \\ -7 \\ \hline \end{array}$	8) $\begin{array}{r} 15 \\ -7 \\ \hline \end{array}$	9) $\begin{array}{r} 3 \\ -1 \\ \hline \end{array}$	10) $\begin{array}{r} 6 \\ -4 \\ \hline \end{array}$
11) $\begin{array}{r} 15 \\ -8 \\ \hline \end{array}$	12) $\begin{array}{r} 14 \\ -7 \\ \hline \end{array}$	13) $\begin{array}{r} 8 \\ -7 \\ \hline \end{array}$	14) $\begin{array}{r} 1 \\ -1 \\ \hline \end{array}$	15) $\begin{array}{r} 9 \\ -4 \\ \hline \end{array}$
16) $\begin{array}{r} 9 \\ -9 \\ \hline \end{array}$	17) $\begin{array}{r} 9 \\ -2 \\ \hline \end{array}$	18) $\begin{array}{r} 15 \\ -9 \\ \hline \end{array}$	19) $\begin{array}{r} 5 \\ -1 \\ \hline \end{array}$	20) $\begin{array}{r} 12 \\ -3 \\ \hline \end{array}$

Name ____________________ Date ____________________

11 Subtraction Facts Minuends up to 18

Subtract.

1) $11 - 5$	2) $14 - 7$	3) $6 - 0$	4) $9 - 1$	5) $12 - 4$
6) $13 - 5$	7) $12 - 9$	8) $10 - 8$	9) $3 - 0$	10) $8 - 6$
11) $5 - 0$	12) $14 - 6$	13) $6 - 6$	14) $9 - 9$	15) $4 - 2$
16) $9 - 3$	17) $2 - 1$	18) $4 - 3$	19) $6 - 4$	20) $11 - 6$

Name ______________________ Date ______________________

11 Subtraction Facts Minuends up to 18

Subtract.

1) 16 − 9	2) 13 − 7	3) 15 − 8	4) 1 − 1	5) 6 − 4
6) 8 − 0	7) 11 − 8	8) 3 − 3	9) 3 − 2	10) 10 − 3
11) 2 − 1	12) 9 − 9	13) 7 − 2	14) 8 − 3	15) 10 − 2
16) 0 − 0	17) 15 − 7	18) 6 − 0	19) 12 − 4	20) 11 − 7

Name ____________________ Date ____________________

11 Subtraction Facts Minuends up to 18

Subtract.

1) 10 − 2	2) 17 − 9	3) 6 − 0	4) 5 − 4	5) 3 − 1
6) 4 − 1	7) 5 − 0	8) 12 − 9	9) 4 − 2	10) 8 − 7
11) 14 − 5	12) 12 − 5	13) 8 − 6	14) 8 − 8	15) 13 − 5
16) 14 − 8	17) 10 − 6	18) 2 − 1	19) 7 − 5	20) 9 − 9

SOLUTIONS

S

Solutions to Problems

Sections 1 - 11

PAGE: 2

1) 1 +0 1	2) 5 +5 10	3) 2 +3 5	4) 3 +0 3	5) 3 +4 7
6) 7 +1 8	7) 2 +5 7	8) 4 +2 6	9) 4 +5 9	10) 0 +4 4
11) 3 +1 4	12) 6 +4 10	13) 4 +3 7	14) 2 +8 10	15) 3 +3 6
16) 2 +6 8	17) 3 +7 10	18) 7 +3 10	19) 6 +3 9	20) 3 +5 8

PAGE: 3

1) 8 +2 10	2) 5 +1 6	3) 0 +6 6	4) 2 +5 7	5) 4 +1 5
6) 4 +0 4	7) 1 +9 10	8) 0 +2 2	9) 6 +3 9	10) 3 +7 10
11) 4 +3 7	12) 5 +3 8	13) 7 +2 9	14) 7 +3 10	15) 3 +3 6
16) 3 +2 5	17) 5 +4 9	18) 5 +2 7	19) 2 +4 6	20) 2 +7 9

PAGE: 4

1) 1 +9 10	2) 4 +1 5	3) 2 +8 10	4) 2 +7 9	5) 3 +7 10
6) 3 +0 3	7) 0 +5 5	8) 0 +8 8	9) 5 +1 6	10) 8 +2 10
11) 1 +4 5	12) 5 +5 10	13) 3 +6 9	14) 5 +3 8	15) 7 +3 10
16) 3 +2 5	17) 5 +2 7	18) 6 +4 10	19) 3 +3 6	20) 6 +3 9

PAGE: 5

1) 4 +5 9	2) 8 +1 9	3) 2 +8 10	4) 4 +4 8	5) 3 +5 8
6) 2 +7 9	7) 1 +7 8	8) 3 +6 9	9) 6 +3 9	10) 3 +0 3
11) 5 +0 5	12) 0 +6 6	13) 3 +3 6	14) 1 +2 3	15) 7 +2 9
16) 2 +2 4	17) 5 +1 6	18) 5 +4 9	19) 6 +2 8	20) 2 +6 8

PAGE: 6

1) 1 +6 7	2) 3 +2 5	3) 2 +3 5	4) 7 +0 7	5) 2 +7 9
6) 8 +2 10	7) 3 +7 10	8) 2 +4 6	9) 0 +2 2	10) 1 +9 10
11) 4 +5 9	12) 5 +2 7	13) 5 +4 9	14) 0 +5 5	15) 6 +2 8
16) 1 +7 8	17) 4 +2 6	18) 2 +2 4	19) 3 +3 6	20) 4 +3 7

PAGE: 7

1) 4 +2 6	2) 3 +4 7	3) 3 +0 3	4) 9 +0 9	5) 2 +7 9
6) 4 +6 10	7) 7 +2 9	8) 4 +5 9	9) 6 +2 8	10) 4 +4 8
11) 3 +2 5	12) 3 +3 6	13) 3 +1 4	14) 6 +0 6	15) 5 +4 9
16) 2 +2 4	17) 1 +2 3	18) 4 +3 7	19) 3 +5 8	20) 1 +8 9

PAGE: 8

1) 0 +8 8	2) 1 +9 10	3) 4 +3 7	4) 3 +1 4	5) 2 +2 4
6) 4 +1 5	7) 5 +4 9	8) 0 +5 5	9) 6 +2 8	10) 3 +4 7
11) 3 +2 5	12) 4 +5 9	13) 6 +4 10	14) 4 +4 8	15) 7 +3 10
16) 3 +6 9	17) 2 +3 5	18) 3 +3 6	19) 5 +2 7	20) 3 +5 8

PAGE: 9

1) 2 +4 6	2) 2 +5 7	3) 8 +2 10	4) 1 +6 7	5) 7 +1 8
6) 8 +0 8	7) 1 +7 8	8) 2 +8 10	9) 2 +3 5	10) 3 +0 3
11) 3 +7 10	12) 0 +2 2	13) 4 +6 10	14) 4 +5 9	15) 5 +4 9
16) 2 +6 8	17) 5 +3 8	18) 3 +6 9	19) 5 +2 7	20) 7 +2 9

PAGE: 10

1) 0 +8 8	2) 7 +2 9	3) 2 +2 4	4) 4 +4 8	5) 6 +0 6
6) 7 +3 10	7) 3 +2 5	8) 1 +5 6	9) 4 +3 7	10) 2 +0 2
11) 2 +5 7	12) 3 +7 10	13) 1 +4 5	14) 2 +7 9	15) 2 +8 10
16) 6 +2 8	17) 3 +4 7	18) 3 +1 4	19) 3 +3 6	20) 6 +4 10

PAGE: 11

1) 3 +5 8	2) 1 +4 5	3) 7 +1 8	4) 4 +5 9	5) 0 +9 9
6) 5 +2 7	7) 2 +3 5	8) 1 +2 3	9) 4 +3 7	10) 0 +1 1
11) 2 +5 7	12) 7 +3 10	13) 2 +8 10	14) 2 +2 4	15) 7 +2 9
16) 0 +3 3	17) 5 +5 10	18) 6 +3 9	19) 3 +6 9	20) 5 +3 8

PAGE: 12

1) 1 +3 4	2) 6 +3 9	3) 6 +2 8	4) 3 +3 6	5) 0 +8 8
6) 7 +2 9	7) 0 +5 5	8) 5 +5 10	9) 1 +5 6	10) 3 +5 8
11) 4 +6 10	12) 5 +0 5	13) 8 +2 10	14) 4 +3 7	15) 5 +2 7
16) 5 +4 9	17) 6 +4 10	18) 4 +4 8	19) 2 +3 5	20) 2 +7 9

PAGE: 14

1) 0+ 7 =7	2) 0+ 5 =5	3) 5+ 2 =7
4) 8+ 1 =9	5) 0+ 3 =3	6) 2+ 7 =9
7) 1+ 2 =3	8) 4+ 1 =5	9) 2+ 5 =7
10) 6+ 4 =10	11) 6+ 2 =8	12) 6+ 3 =9
13) 5+ 3 =8	14) 2+ 4 =6	15) 3+ 6 =9

PAGE: 15

1) 1+ _7_ =8	2) 1+ _2_ =3	3) 3+ _2_ =5
4) 9+ _1_ =10	5) 0+ _1_ =1	6) 2+ _2_ =4
7) 3+ _0_ =3	8) 2+ _6_ =8	9) 6+ _3_ =9
10) 5+ _4_ =9	11) 4+ _5_ =9	12) 3+ _3_ =6
13) 7+ _3_ =10	14) 6+ _2_ =8	15) 3+ _6_ =9

PAGE: 16

1) 1+ _1_ =2	2) 2+ _3_ =5	3) 3+ _1_ =4
4) 2+ _4_ =6	5) 6+ _1_ =7	6) 5+ _4_ =9
7) 4+ _2_ =6	8) 6+ _4_ =10	9) 3+ _4_ =7
10) 5+ _3_ =8	11) 0+ _2_ =2	12) 4+ _4_ =8
13) 4+ _3_ =7	14) 7+ _0_ =7	15) 2+ _7_ =9

PAGE: 17

1) 4+ _2_ =6	2) 2+ _0_ =2	3) 6+ _3_ =9
4) 8+ _1_ =9	5) 5+ _3_ =8	6) 0+ _5_ =5
7) 0+ _9_ =9	8) 5+ _4_ =9	9) 2+ _7_ =9
10) 1+ _1_ =2	11) 6+ _2_ =8	12) 5+ _2_ =7
13) 1+ _7_ =8	14) 5+ _5_ =10	15) 7+ _3_ =10

PAGE: 18

1) 2+ _5_ =7	2) 3+ _1_ =4	3) 7+ _0_ =7
4) 0+ _1_ =1	5) 7+ _3_ =10	6) 1+ _4_ =5
7) 0+ _3_ =3	8) 1+ _2_ =3	9) 3+ _7_ =10
10) 6+ _3_ =9	11) 2+ _7_ =9	12) 2+ _3_ =5
13) 4+ _5_ =9	14) 2+ _4_ =6	15) 7+ _2_ =9

PAGE: 19

1) 5+ _2_ =7	2) 0+ _6_ =6	3) 8+ _0_ =8
4) 9+ _1_ =10	5) 2+ _2_ =4	6) 1+ _6_ =7
7) 4+ _4_ =8	8) 3+ _3_ =6	9) 3+ _4_ =7
10) 6+ _2_ =8	11) 0+ _1_ =1	12) 1+ _5_ =6
13) 5+ _5_ =10	14) 4+ _3_ =7	15) 5+ _4_ =9

PAGE: 20

1) 3+ _0_ =3	2) 0+ _7_ =7	3) 3+ _6_ =9
4) 1+ _3_ =4	5) 4+ _5_ =9	6) 8+ _2_ =10
7) 6+ _2_ =8	8) 7+ _1_ =8	9) 6+ _3_ =9
10) 1+ _2_ =3	11) 4+ _2_ =6	12) 3+ _7_ =10
13) 2+ _4_ =6	14) 3+ _2_ =5	15) 3+ _4_ =7

PAGE: 21

1) 2+ 5 =7	2) 0+ 9 =9	3) 9+ 1 =10
4) 1+ 6 =7	5) 5+ 2 =7	6) 5+ 4 =9
7) 1+ 8 =9	8) 3+ 2 =5	9) 2+ 7 =9
10) 4+ 2 =6	11) 4+ 5 =9	12) 3+ 3 =6
13) 4+ 3 =7	14) 5+ 3 =8	15) 7+ 3 =10

PAGE: 22

1) 2+ 7 =9	2) 7+ 3 =10	3) 3+ 6 =9
4) 3+ 7 =10	5) 4+ 3 =7	6) 2+ 2 =4
7) 1+ 5 =6	8) 6+ 3 =9	9) 2+ 8 =10
10) 3+ 4 =7	11) 8+ 0 =8	12) 0+ 8 =8
13) 4+ 2 =6	14) 4+ 4 =8	15) 3+ 5 =8

PAGE: 23

1) 6+ 4 =10	2) 7+ 1 =8	3) 4+ 2 =6
4) 8+ 2 =10	5) 3+ 7 =10	6) 2+ 8 =10
7) 3+ 1 =4	8) 7+ 0 =7	9) 5+ 2 =7
10) 6+ 3 =9	11) 2+ 6 =8	12) 1+ 8 =9
13) 2+ 7 =9	14) 0+ 8 =8	15) 0+ 2 =2

PAGE: 24

1) 4+ 5 =9	2) 5+ 3 =8	3) 1+ 0 =1
4) 7+ 3 =10	5) 6+ 0 =6	6) 1+ 7 =8
7) 8+ 2 =10	8) 1+ 4 =5	9) 4+ 6 =10
10) 2+ 2 =4	11) 3+ 6 =9	12) 4+ 3 =7
13) 4+ 4 =8	14) 1+ 6 =7	15) 2+ 7 =9

PAGE: 26

1) 10	2) 9	3) 8	4) 3	5) 7
-7	-7	-2	-3	-3
3	2	6	0	4
6) 2	7) 4	8) 10	9) 5	10) 9
-2	-2	-2	-1	-1
0	2	8	4	8
11) 9	12) 6	13) 7	14) 6	15) 4
-3	-2	-2	-3	-3
6	4	5	3	1
16) 8	17) 10	18) 7	19) 8	20) 10
-6	-8	-7	-4	-3
2	2	0	4	7

PAGE: 27

1) 1	2) 10	3) 8	4) 4	5) 8
-0	-7	-2	-4	-4
1	3	6	0	4
6) 2	7) 9	8) 9	9) 5	10) 10
-1	-2	-7	-5	-3
1	7	2	0	7
11) 9	12) 10	13) 5	14) 10	15) 6
-6	-2	-4	-5	-3
3	8	1	5	3
16) 5	17) 4	18) 6	19) 10	20) 8
-2	-2	-2	-6	-6
3	2	4	4	2

PAGE: 28

1) 10 - 9 = 1	2) 8 - 1 = 7	3) 7 - 5 = 2	4) 10 - 6 = 4	5) 8 - 5 = 3
6) 5 - 3 = 2	7) 5 - 0 = 5	8) 10 - 2 = 8	9) 6 - 0 = 6	10) 10 - 4 = 6
11) 8 - 0 = 8	12) 6 - 4 = 2	13) 9 - 3 = 6	14) 5 - 2 = 3	15) 8 - 4 = 4
16) 9 - 6 = 3	17) 10 - 7 = 3	18) 8 - 6 = 2	19) 7 - 3 = 4	20) 10 - 8 = 2

PAGE: 29

1) 5 - 5 = 0	2) 9 - 5 = 4	3) 10 - 6 = 4	4) 7 - 0 = 7	5) 4 - 1 = 3
6) 8 - 2 = 6	7) 8 - 0 = 8	8) 8 - 6 = 2	9) 5 - 3 = 2	10) 5 - 1 = 4
11) 6 - 3 = 3	12) 3 - 2 = 1	13) 10 - 4 = 6	14) 10 - 7 = 3	15) 9 - 6 = 3
16) 10 - 5 = 5	17) 10 - 2 = 8	18) 9 - 4 = 5	19) 4 - 2 = 2	20) 6 - 2 = 4

PAGE: 30

1) 7 - 3 = 4	2) 3 - 2 = 1	3) 9 - 5 = 4	4) 10 - 2 = 8	5) 5 - 5 = 0
6) 7 - 4 = 3	7) 4 - 3 = 1	8) 7 - 6 = 1	9) 7 - 0 = 7	10) 6 - 5 = 1
11) 10 - 4 = 6	12) 6 - 4 = 2	13) 10 - 7 = 3	14) 8 - 6 = 2	15) 9 - 7 = 2
16) 5 - 2 = 3	17) 10 - 8 = 2	18) 8 - 2 = 6	19) 8 - 5 = 3	20) 9 - 2 = 7

PAGE: 31

1) 7 - 7 = 0	2) 4 - 2 = 2	3) 9 - 2 = 7	4) 9 - 7 = 2	5) 5 - 0 = 5
6) 6 - 6 = 0	7) 6 - 2 = 4	8) 7 - 4 = 3	9) 8 - 6 = 2	10) 6 - 1 = 5
11) 5 - 3 = 2	12) 8 - 5 = 3	13) 10 - 4 = 6	14) 10 - 6 = 4	15) 7 - 5 = 2
16) 10 - 1 = 9	17) 10 - 9 = 1	18) 6 - 5 = 1	19) 5 - 2 = 3	20) 10 - 3 = 7

PAGE: 32

1) 6 - 3 = 3	2) 8 - 2 = 6	3) 8 - 6 = 2	4) 4 - 2 = 2	5) 1 - 0 = 1
6) 8 - 5 = 3	7) 7 - 7 = 0	8) 9 - 9 = 0	9) 10 - 8 = 2	10) 10 - 7 = 3
11) 4 - 1 = 3	12) 10 - 2 = 8	13) 9 - 6 = 3	14) 8 - 4 = 4	15) 10 - 3 = 7
16) 5 - 3 = 2	17) 5 - 2 = 3	18) 7 - 4 = 3	19) 10 - 6 = 4	20) 9 - 4 = 5

PAGE: 33

1) 8 - 5 = 3	2) 8 - 4 = 4	3) 7 - 3 = 4	4) 5 - 4 = 1	5) 9 - 9 = 0
6) 10 - 1 = 9	7) 4 - 1 = 3	8) 9 - 5 = 4	9) 8 - 0 = 8	10) 7 - 0 = 7
11) 4 - 2 = 2	12) 7 - 5 = 2	13) 7 - 2 = 5	14) 5 - 2 = 3	15) 7 - 4 = 3
16) 6 - 2 = 4	17) 10 - 2 = 8	18) 9 - 6 = 3	19) 10 - 7 = 3	20) 9 - 4 = 5

PAGE: 34

1) 10 -6 4	2) 3 -3 0	3) 10 -5 5	4) 4 -2 2	5) 6 -5 1
6) 8 -5 3	7) 2 -0 2	8) 7 -6 1	9) 10 -7 3	10) 7 -1 6
11) 9 -5 4	12) 5 -2 3	13) 9 -7 2	14) 8 -2 6	15) 6 -2 4
16) 9 -4 5	17) 9 -6 3	18) 5 -3 2	19) 9 -3 6	20) 9 -2 7

PAGE: 35

1) 6 -2 4	2) 5 -2 3	3) 10 -8 2	4) 5 -0 5	5) 8 -3 5
6) 6 -1 5	7) 6 -4 2	8) 8 -8 0	9) 10 -1 9	10) 4 -0 4
11) 9 -1 8	12) 10 -6 4	13) 9 -6 3	14) 8 -6 2	15) 9 -4 5
16) 9 -3 6	17) 7 -3 4	18) 9 -2 7	19) 4 -2 2	20) 9 -5 4

PAGE: 36

1) 10 -6 4	2) 9 -2 7	3) 9 -7 2	4) 10 -1 9	5) 7 -0 7
6) 3 -3 0	7) 8 -2 6	8) 10 -3 7	9) 7 -1 6	10) 10 -5 5
11) 6 -3 3	12) 7 -5 2	13) 8 -3 5	14) 9 -6 3	15) 6 -1 5
16) 9 -3 6	17) 10 -4 6	18) 10 -7 3	19) 5 -4 1	20) 6 -2 4

PAGE: 38

1)

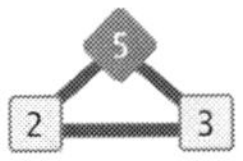

2 + 3 = 5
3 + 2 = 5
5 - 2 = 3
5 - 3 = 2

2)

1 + 7 = 8
7 + 1 = 8
8 - 1 = 7
8 - 7 = 1

3)

1 + 2 = 3
2 + 1 = 3
3 - 1 = 2
3 - 2 = 1

4)

5 + 5 = 10
10 - 5 = 5

PAGE: 39

1)

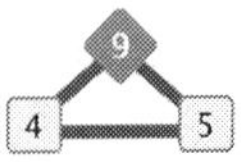

4 + 5 = 9
5 + 4 = 9
9 - 4 = 5
9 - 5 = 4

2)

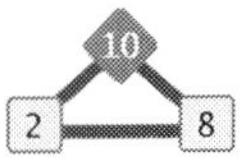

2 + 8 = 10
8 + 2 = 10
10 - 2 = 8
10 - 8 = 2

3)

1 + 3 = 4
3 + 1 = 4
4 - 1 = 3
4 - 3 = 1

4)

3 + 4 = 7
4 + 3 = 7
7 - 3 = 4
7 - 4 = 3

PAGE: 40

1)

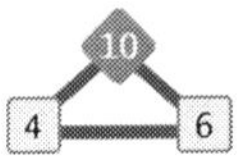

4 + 6 = 10
6 + 4 = 10
10 - 4 = 6
10 - 6 = 4

2)

1 + 4 = 5
4 + 1 = 5
5 - 1 = 4
5 - 4 = 1

3)

2 + 5 = 7
5 + 2 = 7
7 - 2 = 5
7 - 5 = 2

4)

2 + 6 = 8
6 + 2 = 8
8 - 2 = 6
8 - 6 = 2

PAGE: 41

1) 8 — 3, 5

3 + 5 = 8
5 + 3 = 8
8 - 3 = 5
8 - 5 = 3

2) 9 — 2, 7

2 + 7 = 9
7 + 2 = 9
9 - 2 = 7
9 - 7 = 2

3) 6 — 1, 5

1 + 5 = 6
5 + 1 = 6
6 - 1 = 5
6 - 5 = 1

4) 4 — 2, 2

2 + 2 = 4
4 - 2 = 2

PAGE: 42

1) 6 — 2, 4

2 + 4 = 6
4 + 2 = 6
6 - 2 = 4
6 - 4 = 2

2) 8 — 4, 4

4 + 4 = 8
8 - 4 = 4

3) 9 — 3, 6

3 + 6 = 9
6 + 3 = 9
9 - 3 = 6
9 - 6 = 3

4) 10 — 1, 9

1 + 9 = 10
9 + 1 = 10
10 - 1 = 9
10 - 9 = 1

PAGE: 43

1) 7 — 1, 6

1 + 6 = 7
6 + 1 = 7
7 - 1 = 6
7 - 6 = 1

2) 10 — 3, 7

3 + 7 = 10
7 + 3 = 10
10 - 3 = 7
10 - 7 = 3

3) 6 — 3, 3

3 + 3 = 6
6 - 3 = 3

4) 9 — 1, 8

1 + 8 = 9
8 + 1 = 9
9 - 1 = 8
9 - 8 = 1

PAGE: 44

1) 6 — 2, 4

2 + 4 = 6
4 + 2 = 6
6 - 2 = 4
6 - 4 = 2

2) 8 — 3, 5

3 + 5 = 8
5 + 3 = 8
8 - 3 = 5
8 - 5 = 3

3) 9 — 2, 7

2 + 7 = 9
7 + 2 = 9
9 - 2 = 7
9 - 7 = 2

4) 10 — 4, 6

4 + 6 = 10
6 + 4 = 10
10 - 4 = 6
10 - 6 = 4

PAGE: 45

1) 4 — 2, 2

2 + 2 = 4
4 - 2 = 2

2) 7 — 3, 4

3 + 4 = 7
4 + 3 = 7
7 - 3 = 4
7 - 4 = 3

3) 10 — 2, 8

2 + 8 = 10
8 + 2 = 10
10 - 2 = 8
10 - 8 = 2

4) 9 — 1, 8

1 + 8 = 9
8 + 1 = 9
9 - 1 = 8
9 - 8 = 1

PAGE: 46

1) 9 — 4, 5

4 + 5 = 9
5 + 4 = 9
9 - 4 = 5
9 - 5 = 4

2) 4 — 1, 3

1 + 3 = 4
3 + 1 = 4
4 - 1 = 3
4 - 3 = 1

3) 5 — 2, 3

2 + 3 = 5
3 + 2 = 5
5 - 2 = 3
5 - 3 = 2

4) 7 — 1, 6

1 + 6 = 7
6 + 1 = 7
7 - 1 = 6
7 - 6 = 1

PAGE: 47

1) Fact triangle: 6; 1, 5

1 + 5 = 6
5 + 1 = 6
6 - 1 = 5
6 - 5 = 1

2) Fact triangle: 8; 4, 4

4 + 4 = 8
8 - 4 = 4

3) Fact triangle: 10; 1, 9

1 + 9 = 10
9 + 1 = 10
10 - 1 = 9
10 - 9 = 1

4) Fact triangle: 7; 2, 5

2 + 5 = 7
5 + 2 = 7
7 - 2 = 5
7 - 5 = 2

PAGE: 48

1) Fact triangle: 10; 3, 7

3 + 7 = 10
7 + 3 = 10
10 - 3 = 7
10 - 7 = 3

2) Fact triangle: 8; 2, 6

2 + 6 = 8
6 + 2 = 8
8 - 2 = 6
8 - 6 = 2

3) Fact triangle: 5; 1, 4

1 + 4 = 5
4 + 1 = 5
5 - 1 = 4
5 - 4 = 1

4) Fact triangle: 6; 3, 3

3 + 3 = 6
6 - 3 = 3

PAGE: 49

Fact triangle: 8; 1, 7

1 + 7 = 8
7 + 1 = 8
8 - 1 = 7
8 - 7 = 1

Fact triangle: 9; 3, 6

3 + 6 = 9
6 + 3 = 9
9 - 3 = 6
9 - 6 = 3

Fact triangle: 10; 5, 5

5 + 5 = 10
10 - 5 = 5

Fact triangle: 3; 1, 2

1 + 2 = 3
2 + 1 = 3
3 - 1 = 2
3 - 2 = 1

PAGE: 52

1) 2 +9 = 11	2) 9 +5 = 14	3) 4 +9 = 13	4) 7 +5 = 12	5) 8 +4 = 12
6) 5 +8 = 13	7) 4 +7 = 11	8) 6 +6 = 12	9) 7 +9 = 16	10) 3 +9 = 12
11) 9 +2 = 11	12) 3 +8 = 11	13) 8 +5 = 13	14) 8 +7 = 15	15) 9 +4 = 13
16) 7 +4 = 11	17) 8 +8 = 16	18) 5 +7 = 12	19) 9 +9 = 18	20) 9 +7 = 16

PAGE: 53

1) 9 +4 = 13	2) 9 +3 = 12	3) 5 +9 = 14	4) 9 +6 = 15	5) 4 +9 = 13
6) 7 +8 = 15	7) 2 +9 = 11	8) 3 +9 = 12	9) 6 +5 = 11	10) 5 +8 = 13
11) 6 +6 = 12	12) 7 +9 = 16	13) 6 +8 = 14	14) 7 +4 = 11	15) 7 +6 = 13
16) 5 +7 = 12	17) 7 +5 = 12	18) 8 +6 = 14	19) 9 +8 = 17	20) 8 +5 = 13

PAGE: 54

1) 5 +7 = 12	2) 9 +7 = 16	3) 6 +8 = 14	4) 5 +6 = 11	5) 3 +9 = 12
6) 7 +4 = 11	7) 7 +9 = 16	8) 8 +8 = 16	9) 5 +8 = 13	10) 6 +5 = 11
11) 8 +5 = 13	12) 9 +6 = 15	13) 9 +8 = 17	14) 2 +9 = 11	15) 8 +6 = 14
16) 6 +6 = 12	17) 4 +8 = 12	18) 6 +9 = 15	19) 9 +5 = 14	20) 4 +9 = 13

PAGE: 55

1) 6 +6 12	2) 2 +9 11	3) 3 +9 12	4) 4 +7 11	5) 7 +8 15
6) 8 +9 17	7) 4 +9 13	8) 9 +9 18	9) 9 +2 11	10) 9 +6 15
11) 7 +4 11	12) 8 +4 12	13) 9 +5 14	14) 7 +6 13	15) 8 +5 13
16) 8 +3 11	17) 4 +8 12	18) 6 +7 13	19) 3 +8 11	20) 6 +9 15

PAGE: 56

1) 9 +6 15	2) 7 +8 15	3) 8 +7 15	4) 9 +7 16	5) 6 +8 14
6) 5 +6 11	7) 9 +8 17	8) 6 +6 12	9) 4 +8 12	10) 5 +9 14
11) 6 +7 13	12) 7 +9 16	13) 7 +4 11	14) 7 +6 13	15) 9 +2 11
16) 5 +7 12	17) 5 +8 13	18) 6 +9 15	19) 8 +8 16	20) 8 +5 13

PAGE: 57

1) 5 +6 11	2) 6 +8 14	3) 9 +3 12	4) 9 +9 18	5) 2 +9 11
6) 5 +9 14	7) 6 +6 12	8) 6 +7 13	9) 4 +7 11	10) 6 +5 11
11) 4 +8 12	12) 9 +4 13	13) 8 +5 13	14) 7 +7 14	15) 7 +5 12
16) 7 +6 13	17) 7 +8 15	18) 9 +2 11	19) 5 +7 12	20) 8 +7 15

PAGE: 58

1) 8 +7 15	2) 8 +5 13	3) 5 +8 13	4) 9 +8 17	5) 7 +8 15
6) 9 +3 12	7) 6 +6 12	8) 7 +9 16	9) 9 +6 15	10) 6 +7 13
11) 4 +9 13	12) 6 +5 11	13) 2 +9 11	14) 9 +9 18	15) 3 +8 11
16) 7 +4 11	17) 4 +8 12	18) 8 +3 11	19) 8 +9 17	20) 9 +7 16

PAGE: 59

1) 8 +3 11	2) 5 +6 11	3) 9 +5 14	4) 9 +8 17	5) 7 +8 15
6) 7 +7 14	7) 8 +5 13	8) 3 +8 11	9) 9 +7 16	10) 7 +5 12
11) 4 +8 12	12) 8 +9 17	13) 7 +6 13	14) 6 +9 15	15) 3 +9 12
16) 9 +3 12	17) 9 +4 13	18) 6 +8 14	19) 7 +9 16	20) 5 +9 14

PAGE: 60

1) 9 +5 14	2) 9 +9 18	3) 7 +5 12	4) 5 +8 13	5) 8 +7 15
6) 8 +6 14	7) 4 +7 11	8) 6 +7 13	9) 5 +9 14	10) 9 +7 16
11) 5 +6 11	12) 8 +4 12	13) 3 +9 12	14) 6 +6 12	15) 8 +8 16
16) 5 +7 12	17) 4 +8 12	18) 2 +9 11	19) 9 +3 12	20) 4 +9 13

PAGE: 61

1) 4 +8 12	2) 8 +9 17	3) 6 +5 11	4) 8 +3 11	5) 8 +5 13
6) 6 +8 14	7) 6 +7 13	8) 9 +7 16	9) 7 +9 16	10) 8 +7 15
11) 2 +9 11	12) 9 +6 15	13) 5 +6 11	14) 7 +8 15	15) 8 +6 14
16) 7 +7 14	17) 5 +9 14	18) 7 +5 12	19) 8 +4 12	20) 3 +9 12

PAGE: 62

1) 7 +9 16	2) 8 +8 16	3) 2 +9 11	4) 8 +4 12	5) 9 +8 17
6) 9 +9 18	7) 4 +8 12	8) 9 +7 16	9) 7 +4 11	10) 8 +5 13
11) 5 +6 11	12) 7 +8 15	13) 6 +6 12	14) 6 +9 15	15) 6 +8 14
16) 5 +8 13	17) 3 +8 11	18) 4 +7 11	19) 7 +5 12	20) 6 +7 13

PAGE: 64

1) 5+ 6 =11　2) 6+ 5 =11　3) 6+ 8 =14

4) 7+ 4 =11　5) 8+ 8 =16　6) 4+ 9 =13

7) 5+ 9 =14　8) 8+ 5 =13　9) 5+ 8 =13

10) 8+ 7 =15　11) 8+ 4 =12　12) 3+ 8 =11

13) 6+ 9 =15　14) 9+ 9 =18　15) 9+ 8 =17

PAGE: 65

1) 9+ 2 =11　2) 9+ 3 =12　3) 5+ 8 =13

4) 6+ 9 =15　5) 7+ 6 =13　6) 7+ 5 =12

7) 8+ 5 =13　8) 9+ 5 =14　9) 9+ 7 =16

10) 2+ 9 =11　11) 6+ 8 =14　12) 9+ 4 =13

13) 7+ 7 =14　14) 5+ 7 =12　15) 6+ 5 =11

PAGE: 66

1) 2+ 9 =11　2) 7+ 8 =15　3) 8+ 4 =12

4) 8+ 5 =13　5) 6+ 5 =11　6) 7+ 7 =14

7) 3+ 9 =12　8) 6+ 9 =15　9) 5+ 9 =14

10) 5+ 7 =12　11) 9+ 2 =11　12) 9+ 6 =15

13) 8+ 8 =16　14) 8+ 7 =15　15) 6+ 8 =14

PAGE: 67

1) 8+ 7 =15　2) 9+ 5 =14　3) 7+ 8 =15

4) 3+ 8 =11　5) 8+ 8 =16　6) 5+ 8 =13

7) 2+ 9 =11　8) 8+ 9 =17　9) 3+ 9 =12

10) 9+ 7 =16　11) 4+ 7 =11　12) 7+ 4 =11

13) 6+ 8 =14　14) 9+ 3 =12　15) 6+ 9 =15

PAGE: 68

1) 9+ 7 =16 2) 5+ 8 =13 3) 8+ 5 =13

4) 8+ 7 =15 5) 7+ 4 =11 6) 7+ 6 =13

7) 8+ 9 =17 8) 7+ 7 =14 9) 5+ 7 =12

10) 9+ 4 =13 11) 4+ 9 =13 12) 3+ 9 =12

13) 9+ 9 =18 14) 7+ 8 =15 15) 7+ 9 =16

PAGE: 69

1) 5+ 7 =12 2) 2+ 9 =11 3) 4+ 7 =11

4) 8+ 8 =16 5) 8+ 7 =15 6) 6+ 9 =15

7) 8+ 5 =13 8) 7+ 6 =13 9) 9+ 3 =12

10) 8+ 6 =14 11) 8+ 4 =12 12) 4+ 8 =12

13) 6+ 6 =12 14) 9+ 8 =17 15) 3+ 9 =12

PAGE: 70

1) 7+ 9 =16 2) 6+ 7 =13 3) 5+ 6 =11

4) 4+ 9 =13 5) 3+ 9 =12 6) 9+ 5 =14

7) 9+ 4 =13 8) 5+ 9 =14 9) 8+ 7 =15

10) 9+ 6 =15 11) 7+ 7 =14 12) 5+ 7 =12

13) 5+ 8 =13 14) 9+ 2 =11 15) 9+ 8 =17

PAGE: 71

1) 9+ 6 =15 2) 7+ 4 =11 3) 4+ 8 =12

4) 9+ 5 =14 5) 3+ 8 =11 6) 6+ 5 =11

7) 6+ 7 =13 8) 9+ 7 =16 9) 6+ 9 =15

10) 9+ 3 =12 11) 7+ 9 =16 12) 7+ 8 =15

13) 8+ 7 =15 14) 7+ 6 =13 15) 5+ 7 =12

PAGE: 72

1) 4+ 8 =12 2) 4+ 7 =11 3) 9+ 2 =11

4) 8+ 7 =15 5) 6+ 7 =13 6) 5+ 6 =11

7) 5+ 7 =12 8) 8+ 3 =11 9) 3+ 9 =12

10) 4+ 9 =13 11) 8+ 4 =12 12) 9+ 7 =16

13) 6+ 6 =12 14) 3+ 8 =11 15) 8+ 9 =17

PAGE: 73

1) 8+ 6 =14 2) 8+ 3 =11 3) 9+ 3 =12

4) 5+ 7 =12 5) 2+ 9 =11 6) 5+ 8 =13

7) 7+ 8 =15 8) 3+ 8 =11 9) 9+ 9 =18

10) 8+ 4 =12 11) 6+ 7 =13 12) 6+ 9 =15

13) 8+ 9 =17 14) 9+ 6 =15 15) 9+ 7 =16

PAGE: 74

1) 8+ 6 =14	2) 7+ 7 =14	3) 8+ 4 =12
4) 7+ 4 =11	5) 9+ 4 =13	6) 3+ 8 =11
7) 8+ 5 =13	8) 9+ 2 =11	9) 9+ 5 =14
10) 6+ 7 =13	11) 4+ 8 =12	12) 8+ 8 =16
13) 9+ 9 =18	14) 8+ 3 =11	15) 6+ 9 =15

PAGE: 76

1) 11	2) 15	3) 17	4) 11	5) 11
-7	-8	-8	-5	-4
4	7	9	6	7
6) 16	7) 15	8) 14	9) 11	10) 13
-9	-7	-8	-9	-4
7	8	6	2	9
11) 17	12) 12	13) 14	14) 12	15) 12
-9	-6	-7	-5	-4
8	6	7	7	8
16) 12	17) 14	18) 11	19) 16	20) 12
-9	-9	-8	-8	-3
3	5	3	8	9

PAGE: 77

1) 12	2) 12	3) 14	4) 11	5) 14
-9	-5	-9	-3	-8
3	7	5	8	6
6) 11	7) 11	8) 17	9) 12	10) 15
-9	-6	-9	-7	-8
2	5	8	5	7
11) 13	12) 11	13) 17	14) 12	15) 13
-9	-5	-8	-6	-5
4	6	9	6	8
16) 18	17) 12	18) 15	19) 14	20) 11
-9	-8	-9	-5	-8
9	4	6	9	3

PAGE: 78

1) 15	2) 12	3) 15	4) 17	5) 12
-6	-7	-7	-8	-8
9	5	8	9	4
6) 12	7) 12	8) 14	9) 18	10) 17
-4	-5	-6	-9	-9
8	7	8	9	8
11) 16	12) 11	13) 11	14) 13	15) 13
-7	-8	-6	-8	-7
9	3	5	5	6
16) 11	17) 13	18) 14	19) 14	20) 12
-4	-6	-7	-8	-6
7	7	7	6	6

PAGE: 79

1) 14	2) 11	3) 16	4) 16	5) 11
-9	-4	-7	-8	-3
5	7	9	8	8
6) 11	7) 13	8) 12	9) 13	10) 11
-6	-8	-4	-5	-2
5	5	8	8	9
11) 14	12) 18	13) 11	14) 12	15) 11
-7	-9	-5	-8	-9
7	9	6	4	2
16) 15	17) 12	18) 17	19) 12	20) 14
-9	-7	-9	-5	-5
6	5	8	7	9

PAGE: 80

1) 13	2) 12	3) 13	4) 15	5) 12
-7	-3	-8	-7	-5
6	9	5	8	7
6) 14	7) 14	8) 15	9) 11	10) 12
-7	-8	-6	-9	-8
7	6	9	2	4
11) 14	12) 13	13) 11	14) 11	15) 18
-6	-4	-5	-8	-9
8	9	6	3	9
16) 11	17) 15	18) 11	19) 16	20) 16
-3	-8	-4	-7	-8
8	7	7	9	8

PAGE: 81

1) 16	2) 11	3) 14	4) 16	5) 12
-9	-3	-7	-8	-6
7	8	7	8	6
6) 18	7) 16	8) 14	9) 12	10) 13
-9	-7	-6	-4	-4
9	9	8	8	9
11) 13	12) 13	13) 14	14) 14	15) 11
-9	-7	-5	-9	-9
4	6	9	5	2
16) 12	17) 13	18) 15	19) 14	20) 11
-9	-6	-7	-8	-6
3	7	8	6	5

PAGE: 82

1) 11	2) 13	3) 14	4) 16	5) 15
-9	-4	-5	-9	-7
2	9	9	7	8
6) 14	7) 12	8) 13	9) 15	10) 18
-7	-6	-9	-8	-9
7	6	4	7	9
11) 12	12) 12	13) 11	14) 11	15) 14
-9	-3	-8	-2	-9
3	9	3	9	5
16) 12	17) 17	18) 12	19) 11	20) 14
-8	-9	-5	-3	-8
4	8	7	8	6

PAGE: 83

1) 15	2) 12	3) 11	4) 13	5) 12
-7	-9	-2	-4	-6
8	3	9	9	6
6) 12	7) 14	8) 11	9) 15	10) 12
-5	-7	-3	-9	-3
7	7	8	6	9
11) 13	12) 11	13) 16	14) 11	15) 12
-9	-4	-8	-8	-4
4	7	8	3	8
16) 17	17) 12	18) 17	19) 14	20) 13
-9	-8	-8	-9	-7
8	4	9	5	6

PAGE: 84

1) 12	2) 11	3) 12	4) 15	5) 11
-6	-9	-8	-6	-2
6	2	4	9	9
6) 12	7) 11	8) 11	9) 12	10) 13
-7	-5	-7	-9	-6
5	6	4	3	7
11) 12	12) 12	13) 14	14) 14	15) 15
-3	-4	-7	-5	-7
9	8	7	9	8
16) 17	17) 13	18) 11	19) 14	20) 11
-9	-4	-6	-9	-4
8	9	5	5	7

PAGE: 85

1) 13	2) 11	3) 12	4) 17	5) 15
-8	-3	-7	-9	-9
5	8	5	8	6
6) 13	7) 11	8) 14	9) 12	10) 15
-6	-9	-8	-4	-7
7	2	6	8	8
11) 18	12) 11	13) 16	14) 11	15) 15
-9	-6	-9	-7	-6
9	5	7	4	9
16) 17	17) 12	18) 14	19) 11	20) 12
-8	-5	-7	-4	-9
9	7	7	7	3

PAGE: 86

1) 12	2) 11	3) 17	4) 14	5) 11
-4	-6	-9	-7	-4
8	5	8	7	7
6) 15	7) 14	8) 16	9) 11	10) 13
-6	-9	-7	-8	-4
9	5	9	3	9
11) 16	12) 14	13) 11	14) 15	15) 18
-8	-5	-9	-8	-9
8	9	2	7	9
16) 12	17) 14	18) 13	19) 12	20) 15
-7	-8	-8	-9	-7
5	6	5	3	8

PAGE: 88

1)

3 + 9 = 12
9 + 3 = 12
12 - 3 = 9
12 - 9 = 3

2)

8 + 9 = 17
9 + 8 = 17
17 - 8 = 9
17 - 9 = 8

3)

6 + 7 = 13
7 + 6 = 13
13 - 6 = 7
13 - 7 = 6

4)

5 + 6 = 11
6 + 5 = 11
11 - 5 = 6
11 - 6 = 5

PAGE: 89

1)

5 + 9 = 14
9 + 5 = 14
14 - 5 = 9
14 - 9 = 5

2)

7 + 8 = 15
8 + 7 = 15
15 - 7 = 8
15 - 8 = 7

3)

2 + 9 = 11
9 + 2 = 11
11 - 2 = 9
11 - 9 = 2

4)

8 + 8 = 16
16 - 8 = 8

PAGE: 90

1)

6 + 8 = 14
8 + 6 = 14
14 - 6 = 8
14 - 8 = 6

2)

4 + 8 = 12
8 + 4 = 12
12 - 4 = 8
12 - 8 = 4

3)

4 + 7 = 11
7 + 4 = 11
11 - 4 = 7
11 - 7 = 4

4)

9 + 9 = 18
18 - 9 = 9

PAGE: 91

1)

6 + 6 = 12
12 - 6 = 6

2)

7 + 9 = 16
9 + 7 = 16
16 - 7 = 9
16 - 9 = 7

3)

4 + 9 = 13
9 + 4 = 13
13 - 4 = 9
13 - 9 = 4

4)

3 + 8 = 11
8 + 3 = 11
11 - 3 = 8
11 - 8 = 3

PAGE: 92

1)

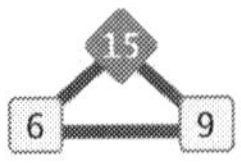

6 + 9 = 15
9 + 6 = 15
15 - 6 = 9
15 - 9 = 6

2)

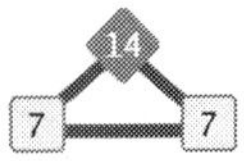

7 + 7 = 14
14 - 7 = 7

3)

5 + 8 = 13
8 + 5 = 13
13 - 5 = 8
13 - 8 = 5

4)

5 + 7 = 12
7 + 5 = 12
12 - 5 = 7
12 - 7 = 5

PAGE: 93

1)

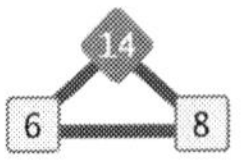

6 + 8 = 14
8 + 6 = 14
14 - 6 = 8
14 - 8 = 6

2)

7 + 8 = 15
8 + 7 = 15
15 - 7 = 8
15 - 8 = 7

3)

7 + 9 = 16
9 + 7 = 16
16 - 7 = 9
16 - 9 = 7

4)

6 + 7 = 13
7 + 6 = 13
13 - 6 = 7
13 - 7 = 6

PAGE: 94

1) 11, 2, 9

2 + 9 = 11
9 + 2 = 11
11 - 2 = 9
11 - 9 = 2

2) 12, 5, 7

5 + 7 = 12
7 + 5 = 12
12 - 5 = 7
12 - 7 = 5

3) 12, 6, 6

6 + 6 = 12
12 - 6 = 6

4) 13, 4, 9

4 + 9 = 13
9 + 4 = 13
13 - 4 = 9
13 - 9 = 4

PAGE: 95

1) 12, 3, 9

3 + 9 = 12
9 + 3 = 12
12 - 3 = 9
12 - 9 = 3

2) 14, 5, 9

5 + 9 = 14
9 + 5 = 14
14 - 5 = 9
14 - 9 = 5

3) 11, 3, 8

3 + 8 = 11
8 + 3 = 11
11 - 3 = 8
11 - 8 = 3

4) 14, 7, 7

7 + 7 = 14
14 - 7 = 7

PAGE: 96

1) 13, 5, 8

5 + 8 = 13
8 + 5 = 13
13 - 5 = 8
13 - 8 = 5

2) 11, 5, 6

5 + 6 = 11
6 + 5 = 11
11 - 5 = 6
11 - 6 = 5

3) 18, 9, 9

9 + 9 = 18
18 - 9 = 9

4) 12, 4, 8

4 + 8 = 12
8 + 4 = 12
12 - 4 = 8
12 - 8 = 4

PAGE: 97

1) 15, 6, 9

6 + 9 = 15
9 + 6 = 15
15 - 6 = 9
15 - 9 = 6

2) 16, 8, 8

8 + 8 = 16
16 - 8 = 8

3) 17, 8, 9

8 + 9 = 17
9 + 8 = 17
17 - 8 = 9
17 - 9 = 8

4) 11, 4, 7

4 + 7 = 11
7 + 4 = 11
11 - 4 = 7
11 - 7 = 4

PAGE: 100

1) 1 +3 4	2) 5 +4 9	3) 3 +8 11	4) 9 +1 10	5) 7 +5 12
6) 4 +7 11	7) 7 +3 10	8) 1 +5 6	9) 1 +9 10	10) 7 +6 13
11) 7 +7 14	12) 9 +7 16	13) 9 +6 15	14) 7 +8 15	15) 8 +0 8
16) 3 +5 8	17) 8 +9 17	18) 5 +3 8	19) 6 +9 15	20) 6 +8 14

PAGE: 101

1) 2 +0 2	2) 6 +9 15	3) 7 +8 15	4) 4 +1 5	5) 9 +2 11
6) 3 +7 10	7) 2 +9 11	8) 1 +5 6	9) 4 +7 11	10) 3 +6 9
11) 4 +5 9	12) 6 +4 10	13) 8 +5 13	14) 0 +4 4	15) 9 +3 12
16) 7 +7 14	17) 0 +7 7	18) 6 +8 14	19) 6 +2 8	20) 2 +3 5

PAGE: 102

1) 7 +1 8	2) 8 +5 13	3) 2 +8 10	4) 7 +3 10	5) 8 +2 10
6) 2 +7 9	7) 9 +7 16	8) 6 +7 13	9) 2 +4 6	10) 6 +5 11
11) 8 +7 15	12) 2 +1 3	13) 9 +9 18	14) 6 +9 15	15) 5 +8 13
16) 1 +9 10	17) 9 +6 15	18) 2 +2 4	19) 3 +9 12	20) 3 +3 6

PAGE: 103

1) 9 +9 18	2) 7 +9 16	3) 0 +3 3	4) 8 +9 17	5) 6 +7 13
6) 2 +5 7	7) 3 +4 7	8) 1 +3 4	9) 7 +4 11	10) 5 +1 6
11) 9 +5 14	12) 7 +1 8	13) 5 +6 11	14) 9 +8 17	15) 4 +3 7
16) 8 +5 13	17) 6 +3 9	18) 8 +4 12	19) 2 +2 4	20) 5 +4 9

PAGE: 104

1) 4 +3 7	2) 7 +8 15	3) 5 +5 10	4) 5 +8 13	5) 6 +6 12
6) 2 +9 11	7) 8 +7 15	8) 0 +8 8	9) 5 +1 6	10) 7 +9 16
11) 1 +1 2	12) 6 +0 6	13) 7 +3 10	14) 6 +1 7	15) 9 +8 17
16) 5 +4 9	17) 5 +9 14	18) 9 +6 15	19) 9 +1 10	20) 9 +9 18

PAGE: 105

1) 1 +5 6	2) 6 +3 9	3) 7 +5 12	4) 0 +3 3	5) 3 +2 5
6) 5 +0 5	7) 6 +8 14	8) 4 +9 13	9) 3 +6 9	10) 3 +5 8
11) 9 +9 18	12) 7 +8 15	13) 4 +1 5	14) 2 +9 11	15) 6 +6 12
16) 4 +7 11	17) 6 +7 13	18) 2 +8 10	19) 1 +9 10	20) 8 +8 16

PAGE: 106

1) 5 +8 13	2) 9 +8 17	3) 3 +7 10	4) 2 +4 6	5) 8 +0 8
6) 3 +9 12	7) 6 +6 12	8) 5 +3 8	9) 7 +5 12	10) 9 +1 10
11) 7 +3 10	12) 1 +9 10	13) 4 +1 5	14) 7 +9 16	15) 3 +4 7
16) 2 +7 9	17) 9 +3 12	18) 5 +7 12	19) 8 +5 13	20) 8 +8 16

PAGE: 107

1) 7 +9 16	2) 3 +6 9	3) 3 +5 8	4) 5 +2 7	5) 6 +8 14
6) 2 +5 7	7) 5 +8 13	8) 7 +2 9	9) 4 +8 12	10) 7 +5 12
11) 5 +6 11	12) 1 +0 1	13) 1 +2 3	14) 9 +4 13	15) 9 +6 15
16) 8 +6 14	17) 6 +5 11	18) 6 +6 12	19) 4 +5 9	20) 2 +3 5

PAGE: 108

1) 1 +4 5	2) 5 +6 11	3) 6 +3 9	4) 9 +2 11	5) 8 +5 13
6) 9 +6 15	7) 0 +2 2	8) 2 +8 10	9) 1 +2 3	10) 3 +7 10
11) 7 +6 13	12) 2 +9 11	13) 4 +4 8	14) 9 +4 13	15) 2 +1 3
16) 7 +2 9	17) 7 +8 15	18) 8 +2 10	19) 7 +7 14	20) 6 +9 15

PAGE: 109

1) 2 +5 7	2) 5 +1 6	3) 7 +7 14	4) 7 +6 13	5) 2 +4 6
6) 2 +3 5	7) 0 +0 0	8) 9 +9 18	9) 5 +6 11	10) 4 +9 13
11) 3 +9 12	12) 1 +4 5	13) 1 +7 8	14) 4 +6 10	15) 8 +6 14
16) 5 +8 13	17) 3 +6 9	18) 6 +7 13	19) 7 +9 16	20) 6 +9 15

PAGE: 110

1) 3 +8 11	2) 8 +4 12	3) 1 +5 6	4) 9 +7 16	5) 3 +9 12
6) 2 +5 7	7) 4 +6 10	8) 5 +3 8	9) 0 +4 4	10) 1 +7 8
11) 6 +4 10	12) 4 +8 12	13) 5 +2 7	14) 9 +4 13	15) 7 +6 13
16) 7 +4 11	17) 9 +2 11	18) 2 +3 5	19) 2 +7 9	20) 3 +2 5

PAGE: 112

1) 5+ 5 =10	2) 3+ 4 =7	3) 2+ 4 =6
4) 8+ 7 =15	5) 9+ 8 =17	6) 9+ 4 =13
7) 3+ 3 =6	8) 5+ 8 =13	9) 1+ 6 =7
10) 8+ 2 =10	11) 2+ 9 =11	12) 8+ 3 =11
13) 3+ 2 =5	14) 9+ 6 =15	15) 7+ 8 =15

PAGE: 113

1) 3+ 8 =11	2) 4+ 4 =8	3) 7+ 1 =8
4) 0+ 6 =6	5) 7+ 5 =12	6) 3+ 9 =12
7) 4+ 8 =12	8) 6+ 9 =15	9) 8+ 8 =16
10) 5+ 4 =9	11) 8+ 9 =17	12) 8+ 2 =10
13) 2+ 9 =11	14) 2+ 3 =5	15) 6+ 8 =14

PAGE: 114

1) 2+ 6 =8	2) 2+ 2 =4	3) 6+ 1 =7
4) 8+ 6 =14	5) 9+ 4 =13	6) 7+ 4 =11
7) 0+ 3 =3	8) 5+ 7 =12	9) 3+ 2 =5
10) 6+ 4 =10	11) 8+ 9 =17	12) 6+ 5 =11
13) 4+ 3 =7	14) 2+ 5 =7	15) 1+ 9 =10

PAGE: 115

1) 5+ 8 =13 2) 4+ 1 =5 3) 6+ 4 =10

4) 3+ 1 =4 5) 9+ 9 =18 6) 0+ 4 =4

7) 6+ 3 =9 8) 7+ 5 =12 9) 8+ 7 =15

10) 9+ 8 =17 11) 5+ 9 =14 12) 7+ 9 =16

13) 6+ 6 =12 14) 2+ 8 =10 15) 6+ 2 =8

PAGE: 116

1) 6+ 8 =14 2) 4+ 8 =12 3) 4+ 2 =6

4) 3+ 1 =4 5) 5+ 4 =9 6) 2+ 5 =7

7) 5+ 1 =6 8) 8+ 8 =16 9) 2+ 1 =3

10) 6+ 7 =13 11) 9+ 2 =11 12) 8+ 3 =11

13) 6+ 4 =10 14) 4+ 6 =10 15) 9+ 7 =16

PAGE: 117

1) 8+ 3 =11 2) 4+ 3 =7 3) 8+ 2 =10

4) 8+ 7 =15 5) 5+ 4 =9 6) 0+ 4 =4

7) 1+ 7 =8 8) 3+ 9 =12 9) 5+ 3 =8

10) 3+ 5 =8 11) 4+ 6 =10 12) 9+ 2 =11

13) 2+ 6 =8 14) 7+ 4 =11 15) 1+ 8 =9

PAGE: 118

1) 7+ 0 =7 2) 3+ 7 =10 3) 7+ 5 =12

4) 4+ 5 =9 5) 4+ 6 =10 6) 8+ 4 =12

7) 5+ 3 =8 8) 1+ 3 =4 9) 3+ 2 =5

10) 3+ 9 =12 11) 1+ 1 =2 12) 3+ 3 =6

13) 9+ 8 =17 14) 6+ 4 =10 15) 8+ 6 =14

PAGE: 119

1) 1+ 7 =8 2) 2+ 1 =3 3) 6+ 8 =14

4) 9+ 5 =14 5) 6+ 2 =8 6) 3+ 6 =9

7) 9+ 3 =12 8) 2+ 4 =6 9) 6+ 4 =10

10) 8+ 1 =9 11) 8+ 0 =8 12) 4+ 8 =12

13) 6+ 3 =9 14) 2+ 6 =8 15) 5+ 6 =11

PAGE: 120

1) 9+ 8 =17 2) 6+ 2 =8 3) 1+ 4 =5

4) 1+ 2 =3 5) 5+ 3 =8 6) 1+ 1 =2

7) 3+ 7 =10 8) 2+ 2 =4 9) 9+ 6 =15

10) 0+ 5 =5 11) 4+ 7 =11 12) 6+ 3 =9

13) 8+ 7 =15 14) 6+ 5 =11 15) 2+ 9 =11

PAGE: 121

1) 3+ 8 =11	2) 2+ 4 =6	3) 0+ 5 =5
4) 4+ 5 =9	5) 9+ 8 =17	6) 3+ 3 =6
7) 8+ 7 =15	8) 8+ 4 =12	9) 4+ 7 =11
10) 1+ 7 =8	11) 9+ 7 =16	12) 2+ 9 =11
13) 6+ 9 =15	14) 5+ 5 =10	15) 3+ 9 =12

PAGE: 122

1) 1+ 6 =7	2) 3+ 2 =5	3) 1+ 3 =4
4) 0+ 1 =1	5) 4+ 6 =10	6) 9+ 7 =16
7) 8+ 8 =16	8) 6+ 9 =15	9) 5+ 3 =8
10) 3+ 5 =8	11) 4+ 8 =12	12) 7+ 2 =9
13) 3+ 3 =6	14) 3+ 8 =11	15) 8+ 4 =12

PAGE: 124

1) 15	2) 11	3) 10	4) 9	5) 6
-6	-3	-1	-2	-2
9	8	9	7	4
6) 12	7) 14	8) 16	9) 8	10) 7
-7	-5	-9	-5	-1
5	9	7	3	6
11) 15	12) 5	13) 3	14) 14	15) 13
-7	-5	-0	-9	-8
8	0	3	5	5
16) 4	17) 8	18) 6	19) 6	20) 13
-2	-0	-1	-6	-5
2	8	5	0	8

PAGE: 125

1) 6	2) 12	3) 12	4) 10	5) 2
-3	-5	-3	-3	-1
3	7	9	7	1
6) 3	7) 10	8) 10	9) 5	10) 4
-0	-2	-1	-4	-1
3	8	9	1	3
11) 13	12) 7	13) 14	14) 5	15) 12
-4	-0	-8	-5	-6
9	7	6	0	6
16) 0	17) 9	18) 13	19) 8	20) 14
-0	-6	-7	-2	-5
0	3	6	6	9

PAGE: 126

1) 14	2) 11	3) 8	4) 10	5) 11
-9	-7	-7	-9	-4
5	4	1	1	7
6) 12	7) 8	8) 5	9) 12	10) 11
-8	-2	-4	-4	-8
4	6	1	8	3
11) 7	12) 2	13) 15	14) 2	15) 18
-7	-0	-6	-2	-9
0	2	9	0	9
16) 16	17) 6	18) 1	19) 12	20) 10
-9	-0	-0	-6	-2
7	6	1	6	8

PAGE: 127

1) 10	2) 13	3) 6	4) 4	5) 9
-3	-4	-3	-2	-6
7	9	3	2	3
6) 9	7) 10	8) 15	9) 16	10) 6
-2	-9	-8	-8	-4
7	1	7	8	2
11) 13	12) 12	13) 5	14) 14	15) 2
-7	-9	-3	-8	-1
6	3	2	6	1
16) 1	17) 15	18) 6	19) 5	20) 7
-0	-9	-5	-2	-6
1	6	1	3	1

PAGE: 128

1) 10 -5 5	2) 8 -6 2	3) 4 -2 2	4) 6 -1 5	5) 11 -7 4
6) 15 -7 8	7) 14 -8 6	8) 3 -1 2	9) 7 -3 4	10) 7 -5 2
11) 17 -9 8	12) 13 -8 5	13) 10 -7 3	14) 8 -3 5	15) 12 -3 9
16) 2 -1 1	17) 14 -5 9	18) 6 -0 6	19) 11 -6 5	20) 11 -9 2

PAGE: 129

1) 7 -5 2	2) 15 -8 7	3) 9 -0 9	4) 14 -8 6	5) 4 -3 1
6) 10 -4 6	7) 16 -7 9	8) 6 -1 5	9) 14 -9 5	10) 13 -5 8
11) 6 -3 3	12) 7 -3 4	13) 5 -5 0	14) 16 -9 7	15) 4 -1 3
16) 3 -0 3	17) 10 -8 2	18) 15 -9 6	19) 17 -9 8	20) 9 -8 1

PAGE: 130

1) 15 -6 9	2) 1 -1 0	3) 10 -6 4	4) 9 -7 2	5) 17 -8 9
6) 2 -0 2	7) 12 -5 7	8) 7 -0 7	9) 6 -1 5	10) 12 -9 3
11) 14 -9 5	12) 9 -6 3	13) 6 -0 6	14) 4 -3 1	15) 17 -9 8
16) 15 -8 7	17) 11 -3 8	18) 13 -8 5	19) 6 -4 2	20) 14 -6 8

PAGE: 131

1) 8 -4 4	2) 13 -5 8	3) 14 -6 8	4) 7 -5 2	5) 2 -0 2
6) 11 -6 5	7) 12 -7 5	8) 15 -7 8	9) 3 -1 2	10) 6 -4 2
11) 15 -8 7	12) 14 -7 7	13) 8 -7 1	14) 1 -1 0	15) 9 -4 5
16) 9 -9 0	17) 9 -2 7	18) 15 -9 6	19) 5 -1 4	20) 12 -3 9

PAGE: 132

1) 11 -5 6	2) 14 -7 7	3) 6 -0 6	4) 9 -1 8	5) 12 -4 8
6) 13 -5 8	7) 12 -9 3	8) 10 -8 2	9) 3 -0 3	10) 8 -6 2
11) 5 -0 5	12) 14 -6 8	13) 6 -6 0	14) 9 -9 0	15) 4 -2 2
16) 9 -3 6	17) 2 -1 1	18) 4 -3 1	19) 6 -4 2	20) 11 -6 5

PAGE: 133

1) 16 -9 7	2) 13 -7 6	3) 15 -8 7	4) 1 -1 0	5) 6 -4 2
6) 8 -0 8	7) 11 -8 3	8) 3 -3 0	9) 3 -2 1	10) 10 -3 7
11) 2 -1 1	12) 9 -9 0	13) 7 -2 5	14) 8 -3 5	15) 10 -2 8
16) 0 -0 0	17) 15 -7 8	18) 6 -0 6	19) 12 -4 8	20) 11 -7 4

PAGE: 134

1) 10 -2 8	2) 17 -9 8	3) 6 -0 6	4) 5 -4 1	5) 3 -1 2
6) 4 -1 3	7) 5 -0 5	8) 12 -9 3	9) 4 -2 2	10) 8 -7 1
11) 14 -5 9	12) 12 -5 7	13) 8 -6 2	14) 8 -8 0	15) 13 -5 8
16) 14 -8 6	17) 10 -6 4	18) 2 -1 1	19) 7 -5 2	20) 9 -9 0

Made in United States
Troutdale, OR
11/03/2026

56847386R00093